Benjamin Franklin

The Autobiography of Benjamin Franklin

(American Polymath and a Founding Father of the United States of America)

Brian Alvarado

Published By **Phil Dawson**

Brian Alvarado

Benjamin Franklin: The Autobiography of Benjamin Franklin (American Polymath and a Founding Father of the United States of America)

ISBN 978-1-9992844-9-7

No part of this guidebook shall be reproduced in any form without permission in writing from the publisher except in the case of brief quotations embodied in critical articles or reviews.

Legal & Disclaimer

The information contained in this book is not designed to replace or take the place of any form of medicine or professional medical advice. The information in this book has been provided for educational & entertainment purposes only.

The information contained in this book has been compiled from sources deemed reliable, and it is accurate to the best of the Author's knowledge; however, the Author cannot guarantee its accuracy and validity and cannot be held liable for any errors or omissions. Changes are periodically made to this book. You must consult your doctor or get professional medical advice before using any of the suggested remedies, techniques, or information in this book.

Upon using the information contained in this book, you agree to hold harmless the Author from and against any damages, costs, and expenses, including any legal fees potentially resulting from the application of any of the information provided by this guide. This disclaimer applies to any damages or injury caused by the use and application, whether directly or indirectly, of any advice or information presented, whether for breach of contract, tort, negligence, personal injury, criminal intent, or under any other cause of action.

You agree to accept all risks of using the information presented inside this book. You need to consult a professional medical practitioner in order to ensure you are both able and healthy enough to participate in this program.

Table Of Contents

Chapter 1: The Childhood Of Benjamin Franklin

Benjamin Franklin was born 17 January 1706 in the home of Josiah and Abiah Franklin, who lived in Boston. Franklin was the 15th of 17 siblings. His father was a migrant from England to the United States in 1682. His mother was a daughter of one of the New England's earliest settlers Peter Folger. Although the Franklin family did not have a lot of money however, they had a good life in their own lives. Josiah owned a candle-making and soap manufacturing business. His mother was busy at the house. She not only cared for all of her small kids, but she cleaned, cooked and sang hymns. She also spun wool, collected her husband's bank records, weaved cloth, sewn and taught children how to pray.

Religion was an essential element within the Franklin family. Benjamin was born in an Puritan family and would spend many Fridays and Sundays watching the music of Cotton Mather while he was a kid. Family members would gather in the church for as long as 4 hours. The Puritans were strict within the church. they required you to stand straight, sit in a straight position and remain silent. In addition you also needed adhere to other regulations like the colour of their clothes and the type of events the church could conduct on Sundays.

When he was a kid, Benjamin spent a lot of time by himself, and didn't get along with his siblings or sisters. In the time in solitude, Benjamin would learn particular life skills, for instance, learning how to swim. The man was able to master these abilities by reading books. Benjamin Franklin loved to read books. He saved every dime he had to purchase books, and borrowed from his friends.

The importance of education was not just within the Franklin family. There was however, not much formal schooling. However, because Josiah was awestruck by the idea of how Benjamin might one day be an Clergyman, Benjamin was enrolled at Boston Grammar School. Boston Grammar School when he was about eighteen years old. The fact is, Benjamin's time in Boston Grammar School Boston Grammar School was short but a few years later Josiah was the one to take Ben from the school. There are many historians who continue to question what the motive was. Josiah took Benjamin out of the school because Benjamin was on the upper end in his class. Many believe the cause was financial, while some believe that Josiah had given up on the possibility of Benjamin becoming the next

Clergyman. Others believe Josiah believed that he could obtain higher education at a different school. No matter what the motivation, Benjamin was later enrolled at the Mr. Brownell's School of Writing as well as Arithmetic.

Then, later in his life Benjamin noted that he was a pro when writing, yet struggled with the area of arithmetic. He was exiled from Mr. Brownell's school for writing as well as Arithmetic within a year. It was the final phase of Benjamin's formal training. His father was of the opinion that Benjamin could learn from home, and also within his soap and candle business. Although Benjamin was not a fan of crafting candles, he enjoyed taking in the life and political advice and advice the father frequently gave to the people of the community. The family would typically visit Josiah as the entire family was seated around the dining table. Josiah allowed

Benjamin to remain at the table and listen to the conversation and expand his mind. In later years, Benjamin wrote that he became so engaged with the conversation, the man would not eat.

BEN FRANKLIN: THE BEGINNING PIECES

When Benjamin was just twelve when he was twelve, his father made the decision to have him become an apprentice for James, his brother James the printing press. It was among Benjamin's first actions in his early life which would eventually lead to his future. Benjamin was the one responsible for making the letters for the printer and also selling newspapers around the town. The job however did not please Benjamin Franklin. In fact, Benjamin wanted to contribute articles for a newspaper rather than just continuing to do his busy schedule. The problem was, Benjamin knew his brother was not a fan of Benjamin writing for a paper, which was dubbed, The Courant. Thus, one morning Benjamin dropped an unidentified article in front at the printing house where his brother was. Instead of signing the article with his own name, Benjamin spelled the article as Silence Dogood. James loved the Dogood letters to the point that the author published them all in the period from the month of April through October 1722. The letters, however, didn't go over well with the Assembly which demanded James to disclose the identity of the writer. If James did not agree, mainly because he didn't know who the true author was and was subsequently put in prison for two weeks. This left Benjamin as the head of the printing company. After James was released, he was barred from printing newspapers, which is why Benjamin was allowed to continue for a few days. But he became tired of the abuse his brother was causing him and chose to use advantage of the chance to get rid of his apprenticeship contract that he made three years prior, when he was twelve.

However, because James went around Boston and urged other printers not to employ Benjamin and he had no chance to secure another print job at Boston. Thus, Ben snuck out of the house of his brother in 1723 at the age of 16 year old, and got on an ocean vessel bound for New York City. It didn't take Benjamin time to realise that he wasn't going to get an employment in printing within New York City. Then, following the guidance given by New York City printers, He set out to the most populous city of the American colonies: Philadelphia.

In Philadelphia, Franklin found a position working with Samuel Keimer. Then, it wasn't much more than the governor of Pennsylvania was able to visit Franklin. Franklin ended up offering Franklin the chance to start the printing company of his own. Benjamin Franklin agreed, and Governor Scott took Franklin to London to purchase materials. But, when Benjamin was in London and realized that there was no way to get him out. The Governor, Sir William Keith did not send the documents of credit or even money. Benjamin was utterly broke as well as homeless and living in a foreign nation. He did the one thing that he was able to do get - locate a print job. In the end, Franklin was hired at Palmers. The company he worked for was for one year and a half, until Thomas Denham paid for Benjamin's return trip to Philadelphia. When Benjamin returned to American ground, he had more about printing than other printers that was in Pennsylvania. In the year following his return returned, Benjamin opened his own printing firm.

Chapter 2: The Beginning Of A Master Mind

THE START OF FREEDOM

Printer for currency printed on paper printing paper currency in New Jersey.

But, holding even one of the top printing positions during Colonial America was not Benjamin's greatest business achievement. Benjamin's greatest achievement in business is his Poor Richard's Almanac. The first edition was published by the author Richard Saunders on December 19 1732. In the following 25 years and beyond, The Poor Richard's Almanac would be released every year. It contained a wide range of data, including meteorological forecasts and demographics, as well as poetry, calendars and proverbs, trivia, recipes, as well as other fascinating details and facts. The Almanac quickly was the most popular resource throughout America. American colonies, particularly for those with no money to buy the cost of books. The almanac was printed in 10,000 copies. Poor Richard's Almanac were printed every year.

Through the popularity of the Poor Richard's Almanac and his numerous business ventures Benjamin Franklin had become very popular and wealthy. However, Benjamin was aware that he would like to take on other responsibilities within his personal life. Therefore, in 1748 Benjamin hired a collaborator to handle the everyday printing needs. Since he was working with David Hall employed, Franklin was able to take part in public and experimental projects. Hall as well as Franklin continued to be printing together until 1766, at which point Franklin was able to sell the company to Hall.

One of Benjamin Franklin's most favorite tasks was to write. Through his career as a writer, he was not just one of the most skilled printers of Colonial America but also became an acclaimed writer across all over the world. He not only wrote hundreds of letters to different people, like family as well as friends and colleagues He also composed and published almanacs, pamphlets or essays.

Apart from The Poor Richard's Almanac, one of Franklin's best-known books included Father Abraham and "The Way to Wealth." Father Abraham was written in the time that Benjamin Franklin was gathering all the information included in the Almanac regarding investing funds. In the articles Benjamin created a short story on Father Abraham, who had been a frequent fan in the Poor Richard's Almanac for 25 years. The story first appeared in the almanac from 1758, and instantly became popular for its audience. While the tale is humorous, it was amusing and kept people laughing.

Benjamin had also been spending many nights creating his autobiography. The autobiography was first started by Benjamin in the latter part of 1760, but could never complete it. As it was time for the American Revolutionary War began, his autobiography was forced to be put aside. autobiography down, and almost went unnoticed. Following his retirement, he began to some work, but only minimally. After his death the autobiography only ran up to 1757. However, despite the fact that it's not yet finished The autobiography written by Benjamin Franklin is one of the most acclaimed works in American literary history.

FAMILY LIFE WITH BENJAMIN FRANKLIN

A RELATIONSHIP OF TWO COLONIALS

Benjamin Franklin got married to his long-time partner, Deborah Read, when

Franklin first arrived in Philadelphia. The couple stayed together for a while Deborah as well as her parents. They soon became lovers, but they had a rocky affair. As Franklin was in England for 18 months, Deborah got married to John Rogers, who was the potter of Philadelphia. Sadly, Deborah and John's happy union was not long-lasting. The couple did not only let John have a frenzied spending habit and incur a large amount of debt and a lot of debt, there was the possibility of him having one wife who was from England. Deborah couldn't take more suffering from John He was a jerk, and she walked to the house of her mother then John went to The West Indies.

In the same time when John left, Franklin returned to the colonies. Benjamin was devastated by the news about their wedding, as the couple could not be married Deborah. It wasn't that long after the rumors began spreading about John was drowned in the sea. The two Deborah and Benjamin wanted to know if these stories could be accurate.

However, Benjamin and Deborah had an unmarried marriage. Deborah began living with Benjamin and started calling her in the form of Mrs. Benjamin Franklin. A common law wedding was a common choice in the 17th century. Deborah and Benjamin thought that this was their ideal wedding as they could not be able to tell whether John was actually dead or an unsubstantiated speculation. If John did not exist, Deborah was not free to wed. Should Deborah and Benjamin have recorded their wedding to the church, and John came back to the church, both Benjamin as well as Deborah could be accused of bigamy. It was a serious criminal act during Colonial America, and it could be punished with life imprisonment as well as 39 lashes.

Through the marriage of Benjamin Franklin, you will see how Benjamin loved and took care of Deborah regardless of the fact that the latter was never in her presence. In their relationship, and throughout the remainder of Benjamin Franklin's lifetime, Benjamin did not speak and write about the positive aspects of his wife's life. Benjamin never criticized Deborah or discussed her weaknesses, even though numerous other individuals Benjamin talked to and wrote to had a sense of her weaknesses. Benjamin did not write about his frustrations with Deborah.

But even if Benjamin and Deborah have a love for each other and praised each other, doesn't necessarily mean that Benjamin and Deborah were in a perfect union. In the time that Benjamin Franklin started to make his mark on the American colonial world by being writer, printer, stateman, scientist and politician; they began drifting away. As more Benjamin was away, being in time with the general public, but the woman he was married to, Deborah was becoming increasingly jealous and lonely.

Pappy was the name Deborah chose for Benjamin.

When Benjamin Franklin's popularity was growing all over the globe, he began traveling around the world. He left in 1757 to England. However much Deborah tried to convince her husband, Benjamin would not return to his home. Benjamin was not even able to return to his home after his wife had suffered a stroke of severe severity, which left her speech impaired as well as a weak memory. This happened from 1768 to 1769. Benjamin Franklin learned of the death of his wife in 1775's February in his newspaper, the

Pennsylvania Gazette, which read, "On Monday, the 19th of December, died Mrs. Deborah Franklin, wife of Dr. Franklin." Although this could seem to make Benjamin appear to have lost hope of a marriage to Deborah however the contrary is evident in his writings shortly after the loss of his wife.

It is a proof point that Benjamin was still a lover and admirer of Deborah the way the way he has always.

FATHER AND GRANDFATHER

Benjamin Franklin's son who was born first was given the name William "Billy" Franklin. It is not known for certain the exact date William was born, but we do can be sure the fact that his birth date was the latter part of 1730 or in early 1731. There was no doubt that many residents of Philadelphia were discussing and discussing rumors concerning Benjamin

Franklin's son who was not his. There were many people, including historians of today, were wondering what the mother of William was. For most, there was only two possibilities. One was that William was Deborah's son. born prior to their common law wedding. The growing sexism which Deborah did to William is often a reason to believe that this wasn't an actual fact. Another option was a woman believed by many to be was a servant within the Franklin family. The woman, named Barbara and was William's mom.

Although we haven't received a specific answer from any one on what William's mother's name is Later in his life Benjamin Franklin wrote that William was born of passion. The passion, Benjamin would explain, was an impulse that he could not manage. But, Benjamin loved William, like he did his children and other siblings, and was fully responsible for the role of being

William's father. Deborah was willing to rear William like her own but, while she did so her best, she showed him very little respect in comparison to her two other children. As the more old William was older, the more Deborah discussed her displeasure with William. Happily for William was a great friend, he got lots of admiration and affection from Benjamin. Although their affection and love to each other would last throughout their lives, they might have a difficult time staying on greatest of terms, especially following they had gone through the American Revolution.

Due to the influence of King George III over William during 1757, when William took his father's place in England, William remained loyal to the monarch during his time in the American Revolution. In America's struggle for freedom the battle was between father and son. Benjamin has fought for his life to protect his life and the and the lives of his fellow Founding Fathers, as well as achieve the independence of America. William was skeptical that America would succeed in the war, or manage itself and was a fervent opponent to the American fight. In the aftermath of during the American Revolution, the two sent letters to one another. The letters proved that they believed in and cherished the other, yet they wouldn't have the ability to transcend the emotions triggered by their differences throughout the American Revolution.

William the writer, who lived in London and was a resident of London, addressed his father with the following message:

"How I would like to rekindle the love affair and bond that until the beginning of the recent troubles been my pride and joy in my life. The reason I did what I did was out of a deep conviction about what I thought was my obligation towards my king and nation. If that was the case in the future, my actions was exactly similar...All that said, this is it is his history...My greatest wish is to resumption our

To Benjamin's note, Benjamin wrote:

"Nothing that has affected me as deeply and caused me intense feelings as be abandoned at the age of 87 by my son. It wasn't just that I was deserted however, to discover his armies advancing at me for a matter that my fame or fortune as well as life was in danger... It is true that there are legitimate obligations that flow from to those of a political nature, and can't be eliminated through them...This is a thorny topic. I am dropping it...I would be delighted to visit you whenever it is you are able, but will not like to see you to me at the moment."

In 1785, when Benjamin made this note to William He was living in France in a state of anger over the deceit of his son who was the first to be born. In 1785, as Benjamin was returning towards America from France He reconnected in England with William on the way to England. It was, however, not the most pleasant father-son reunion. Benjamin couldn't get over his anger towards William. Additionally, William could sense his father's disdain, which made him feel uneasy and uncomfortable. It was the final time William Benjamin and Benjamin will see one another since there was no resolution at the meeting.

William wouldn't be Benjamin Franklin's sole son, but was his sole son who would live to be an adult. 1732 was the year that Benjamin as well as his wife were welcomed by Francis Folger. Francis suffered from smallpox and died after four years of age There isn't much about his past. We do know that he was a fan of his father's footsteps in the printing company. There is also a rumor that Benjamin did not get over the passing of Francis. To the day he died, Benjamin would cry for his son who was the second to be born. Benjamin pondered what could be the fate of Francis. Every time he saw a little child, his thoughts was always occupied by the boy he had lost.

Benjamin Franklin would write of Francis towards the end of his existence, 50 years later after his death.

It was 1743 when Benjamin as well as Deborah would have another baby in their family. At this time they were expecting the birth of a daughter. They were to name Sarah. She was however, often referred to by her nickname, Sally. Since William an teen and Francis had passed on, Sally often felt as like she was the only child.

Since Sally's passion is music Benjamin gave her a piano when he lived in England. When he returned to America The father and daughter spent time with their beloved time creating music. Sally was a harpsichord player while Benjamin was playing the glass harmonica, which was an instrument he created.

Sally was married to a man an initials of Richard Bache in 1767, during the time that Benjamin was still in England. Sally and Richard continued to reside within the Franklin household where they raised their seven children. Sally was also able to care for her father as he grew older years, particularly in the final 4 to 5 years of his life.

Benjamin was unable to come up with a better method to spend his last days than together with his daughters, son-inlaw and his grandkids, or the prattlers in his small gang, which he calls them.

VA KITE, ELECTRICITY, AND INVENTIONS

Chapter 3: Beginning Stages

As a young man his mind was in motion. Benjamin would engage in every day activities like swimming and come up with an idea that of his own to play with. As an example, Benjamin was able to take swimming while fly-flying his kite. He sat on his back, sat in the water, swam with the stick while letting the kite take his body across the pool. In the present, this is thought to be the first Benjamin Franklin-inspired science experiment. Later on in his life he planned to extend his experiment into his home in the English Channel. But, he discovered that it was more secure and safe to utilize boats or ships instead of testing his small idea.

A different activity in his childhood was the creation of his own"square" of squares. It was about forming the square with numbers, so that the total of each vertical, horizontal and diagonal row would be equal. This was the pattern that would guide Benjamin through adulthood, when Benjamin would usually make a magical square out of squares whenever he felt bored or had some time to himself.

Through his early years and even into the age of adulthood, Benjamin Franklin continued to look around, investigate, and scrutinize every aspect of his surroundings. Franklin was curious about how the ants could always get into the molasses jar. He even moved the containers, and each time they would be able to find the ways. Through the various tests he conducted using the jar of molasses and the ants Benjamin believes that ants speak their own language. the moment one of them learns about something like food, it will go back to the colony and inform others insects.

AND ELECTRIFIED LIFE

When Benjamin carried out a variety of experiments, the man is famous for his experiments that involved electricity. Actually the subject of electricity was one of Benjamin Franklin's most popular subjects for experiments. In addition, Benjamin would become famous during his experiments with electricity.

He wrote once "my house was continually full of persons who came to see the new wonders."

Most often, the experiments Benjamin performed before an audience was something he'd performed numerous times. One of these experiments was known as fake spiders. This experiment Benjamin builds a realistic yet fake spider from an unburned cork and a linen thread. Cork served as the body of the fake spider, while linen thread were an imitation spider's legs. The fake spider was tied to the wire. It would spin and move whenever the wire was contacted with electricity that was a battery. This was the brainchild of Benjamin Franklin. Benjamin was always fascinated by this idea since his viewers would think that it was real since it was moving quickly and not know what was real from.

Every time an experiment would be conducted, Benjamin would record the processes and steps and then share the results with his friend Peter Collinson. Peter loved reading about Benjamin's experiments, and advised Franklin that they should be made available for sharing. So, in 1751 Collinson collated all his descriptions of Benjamin Franklin's research and put them into the form of a book. This book, which was called, Experiments and Observations on Electricity, which would eventually make Benjamin Franklin a worldwide scientist.

Everyone has had a look at a picture that depicts Benjamin Franklin and his son, William, flying a kite during a storm, looking for lightning to strike. There are several variants of this painting. It it is one of Benjamin Franklin's most widely debated research.

The event that was the most important occurred in the month of June 1752. The afternoon was when Benjamin was looking out for a powerful storm which was expected to sweep across the area. At at this time, Benjamin had come to believe that lightning and electricity are identical. In order to test this theory, Benjamin made a kite made of silk, instead of paper. The kite was also equipped with the appearance of a rod made of metal in its center. Benjamin was also able to attach an untidy handkerchief to the top to ensure that the person who held the kite could be secured from an electric shock.

The wire that ran through the middle of the rod, and downwards. On the lower end of the wire was there was a key. The Key, Benjamin believed, would allow the collection of an electric charge before moving because of the charges. Benjamin's intention was to fly his kite through the eye of the storm. While the kite was flying during the storm Benjamin hit the key. He was exuberant because he felt an electrical shock that was very familiar. Actually, it was the same kind of shock he received when he was able to touch one of his own invented electrical batteries. In contrast to his previous research, Benjamin did not release the results of this one with scientists immediately. Actually, Benjamin did not even keep a record of the date as well as the time of the test. Benjamin Franklin kept his findings of the kite study in secret for several many months. Even to this day the historians do not know the reason he kept it secret and some speculate that it was due to the fact that he was aware of what this kind of research could bring to the world of science, and was right at the center of this. Some believe that Franklin determined to confirm that he actually discovered his claims to having found.

Benjamin Franklin's kite-making fame as well as lightning experiment inspired him to create the lightning rod.

DISCOVERIES OF ELECTRICITY

In all his work in the field of electricity Benjamin was able to make several observations in his research, and he recorded them. One of those observations was the fact that electricity as well as lightning produce a loud sound when they ignite. Also, he noted that both release the light that can set off fire and be fatal to.

The fact the possibility that electricity and lightning could cause damage to property or even kill someone is one of the main reasons people remained cautious in putting a lighting rod at the top of their houses. Although some did adhere to Benjamin Franklin's suggestions for the lightning rod following their reading about it in Poor Richard's Almanac, many individuals remained convinced that the idea was risky.

Benjamin created the terms "negative" and "positive" in relation to electric charge. Additionally Benjamin also found out that certain substances had superior circumstances than others, thus permitting electricity to flow through them with ease. As part of an earlier experiment prior to the kite's famed study, Benjamin discovered that some material, including silk, are shields from electrical shocks.

Though the majority of his science-related discoveries and research are considered to be minor or non-existent nowadays, they were incredibly significant and important in the years of Franklin's lifetime. They were so innovative in their time that The Royal Society awarded him the Copley Medal, which was the highest award in England for scientists. In addition there were three top institutions, Oxford, Harvard, and Yale have awarded Benjamin Franklin with honorary degrees. That's for some to begin calling Benjamin Franklin as Dr. Franklin. The honorary degrees are some of the most prized possessions which Benjamin Franklin ever received. Franklin was often uncomfortable that he had only 2 years' formal training. In addition to this, even though schooling was less formal rather than formal in Franklin's time however, it wasn't typically that someone like Franklin having so very little formal training, would be so revered in the field of study, and especially those in the scientific world.

AN INVENTOR BEFORE HIS TIME

It wasn't just that Benjamin Franklin dive into scientific research, he was also the inventor of a variety of items. In addition to the armonica made of glass as well as Bifocals.

Benjamin Franklin is well known for his glasses. Although not all his portraits show him wearing glasses however, there are a few that have them. Although most people have heard that the glasses he used however, there is little known fact that, from 1780s to the close of his existence, Benjamin Franklin wore one, at least the initial pair of Bifocals. Benjamin used glasses all of his adulthood, but as time passed the need for two glasses. The glasses he needed were to view these things in close-up view and required a second pair to ensure it could view things further from him. That means Benjamin Franklin was both near and far. This meant that he was required to wear two sets of glasses. At some point, Benjamin felt changing between the two glasses would be difficult. Then, he took both pairs of glasses to a shop. There, he requested the shop owner to cut both pairs of lenses into two halves. He then put the half from the glasses for new lenses. It was then possible to or look at the top portion of his glasses, or the bottom part of his lenses to find the things he wanted for better vision.

To carry on his experiments in science particularly those that were focused on electrical charges, Benjamin realized he needed an apparatus that could provide him with enough electrical charge. As he was aware of the fact that pressing two things in a row could cause tension, Benjamin recognized he had to create something based on this concept. He came up with an electrostatic device. It was an early prototype of an electric generator. It was an elongated glass ball at its top, a rotating wheel as well as a the skin of chamois. If the globe of glass was pressed against the skin of the chamois static electrical charge would be generated. The charge of electricity came from knitting needles that were stored in the Leyden container. Benjamin was pleased that the invention was able to work for his experiments. Thanks to this invention, Benjamin was confident that he would get sufficient electricity to conduct his research.

In his ingenuity, Benjamin did not always come up with an idea. One time, Benjamin tried to create an alphabet that was easier to read to be used in to be used in the English language. It was because Benjamin was shocked by how badly women and especially men were able to spell. Naturally, in Benjamin's time the formal schooling system was not as simple like it is now, women received a bare or very low levels of formal training. Due to the absence of formal education, the people wrote words by the way they sound. Benjamin's alphabet Benjamin took out alphabets that were not compatible with their sound, like the letters q, x and letters like y, q, and x. Benjamin also decided to develop new letters believed were necessary for people to spell more easily in the event that they spell based on what a word's sound. Benjamin experimented with this new alphabet on a number of his acquaintances when writing them letters. Some of them tried to reply in the same manner, it was difficult to everyone, even Benjamin Franklin, to let leave the old alphabet that they'd known for the entirety of their lives. So, he put his own alphabet aside, and continued to use the standard English alphabet. In the same way, Benjamin stated, he could never "uz a simplrverzun al hizlyf."

Chapter 4: Other Pieces Of Benjamin Franklin

A WELL-ROUNDED INDIVIDUAL

It was not just that Benjamin Franklin a scientist and inventor, he also had many other accomplishments throughout his lifetime. One of them was in the form of starting a volunteer fire department located in Philadelphia. Franklin also assisted in the construction of hospitals and was involved in improving formal education. In the time that America was in the midst of 13 colonies at the time, Franklin was its head postmaster. In spite of all these piece of puzzle which contributed to develop the life of Benjamin Franklin, his focus was on being the most ideal person he could be.

Like Thomas Jefferson and many other individuals of the same time, Benjamin Franklin followed an exact daily routine. The reason for this was not just in order to ensure he was staying in the right direction and on track, but also for him to assist himself to become a better person. Benjamin had reached his mid- to late 20s at the time he decided that to strive to be the perfect person. In order to achieve his goal He developed thirteen guidelines to adhere to and wrote them down in a notebook was carried everywhere the go. One of his ideas to get better acquainted using these 13 guidelines was to focus on each week one rule. After week thirteen was over then he'd start from scratch. But, Benjamin eventually came to realise that perfection was not achievable. However, he was able to keep the notebook in his pocket throughout his existence. Not only did he carry it everywhere was he traveled, but he would also display the notebook to explain the thirteen guidelines. The thirteen rules of Benjamin Franklin were in the following order:

1. Temperance. Consume food not until it dulls. Do not drink to height.

2. Silence. Do not speak unless it will help others or you. Do not engage in conversation that is merely superficial.

3. Order. Make sure that everything has its locations. Each aspect of your company have its time.

4. Resolution. Make a commitment to do the things you must. Make sure you do the things you have resolved to do.

5. Frugality. Do everything you can to be a good neighbor to someone else and yourself. i.e. don't waste a dime.

6. Industry. Lose no time. You should always be engaged to do something productive.

7. Sincerity. Use no hurtful deceit. Be honest and sane If you do speak, then speak in a respectful manner.

8. Justice. No wrong, not by causing damage or neglecting the benefits you are entitled to.

9. Moderation. Avoid extremes. Do not be so angry about injuries that you believe they are not worth.

10. Tranquility. Don't be disturbed by a few stray crumbs or in the event of accidents which are commonplace or inevitable.

11. Cleanliness. Accept uncleanness of clothing, your body, or in your home.

12. Chastity. Very rarely use venery, but only in case of health, or to help offspring.

13. Humility. Imitate Jesus and Socrates.

One thing in which Benjamin Franklin continued throughout his time was his volunteerism in the community. Benjamin was a firm believer in"the "best service to God, is doing good to man."

Thus one of the first things Benjamin was likely to have him ask in the beginning of the day would be "What good shall I do today?"

This portion of the day was so crucial to him, he even wrote it in his schedule for the day. On the subject that he was assigned, he had written:

TO BETTER A COMMUNITY IN THE EYES OF FRANKLIN

In 1736, Benjamin Franklin organized the Union Fire Company in 1736, which was to be the city's first department of firefighters that were volunteer. Benjamin had the idea from the club that was established in London and was formed to combat flames. Benjamin assisted in forming an organization of thirty members who appointed Benjamin Franklin as their initial fire chief. They were all not solely trained to tackle flames but also helped in the rescue of people. Because of Benjamin's involvement in the department's firefighter's unit and their monthly meeting on ways to reduce fires The department grew in terms of volunteer numbers.

Another method Benjamin was concerned about his own community's safety is to ensure that the streets were safer. In the same year, he created the department of fire, he made sure that the garbage was removed from the streets as well as enhancing the street lights. They had holes on the top. So instead of letting the flame of the candle inside a globe similar to the old street light and the flame of the candle would escape the opening. In addition, Benjamin hired a group of people to serve as lamps to the city. They were responsible of lighting the candle each the night, and also any maintenance.

One of Benjamin Franklin's objectives throughout his life was to develop higher levels of formal education for boys. Benjamin was always unhappy about not receiving adequate formal education throughout his youth, and was determined to improve the education of his next generation. So, in the 1740s Benjamin published his "Proposals relating to the Education of Youth in Pennsylvania." This publication explained Benjamin Franklin's plans for an education that was free for the boys of the colony. The program was limited to children aged eight to 16 but did not cover girls.

Some historians think that girls weren't included since it wasn't usual for girls to get an official education at the time of Benjamin's time Some are of a different opinion. Although Benjamin Franklin was known to be an incredibly generous person and a good man, he also had a specific perception of women that was somewhat like the people of his time. However advanced in time Benjamin was research and inventing his belief was that women were in some kind of place in their lives, and could not be removed from it. This place, according to Benjamin Franklin's thoughts is the house, which Franklin argued when he described his ideal female:

For a deeper explanation of why historians think Benjamin didn't include girls in his educational plan study the recorded conversation Benjamin held during a conversation with one of his female colleagues and she spoke of her desire to study.

Simply put, Benjamin didn't understand women who needed more than what Benjamin, and many others in his time believed women required. He did, however, believe that women ought to be taught the fundamental skills and functions that could be beneficial for them and their families including writing as well as accounting, reading, and writing but he didn't believe they had to be taught out of the confines of this box. This belief in education was practiced in his home. As his son William took part in every kind of class including history, geography and philosophy. Sally learnt the art of embroidering, knitting and spin.

In addition to the fact that he excluded girls from his official educational strategy, Benjamin Franklin had a extremely clear idea of his school. It was to provide a warm, comfortable of family-like setting for pupils. In the booklet that he wrote about the school Benjamin said that the school must include a garden as well as fields. Benjamin also said that the school should be equipped with a huge library that would include not just books but globes, maps and instruments for experimentation. Also, he wanted swimming to be an integral part of the curriculum at school, in addition to other forms of fitness. Along with presenting the curriculum he wanted to be taught in the school and examining how teachers conduct themselves. This was extremely important for Benjamin since, at one school there was a teacher was not a good fit with, and this negatively affected his academic performance. When he had the opportunity to change teachers and one he was comfortable with as well, he became one of the top students in the class. Benjamin stated that teachers in his school must be patient with children's studying. They should also demonstrate moral values and act as the model for these morals.

When Benjamin issued his guide to the school's mission it was founded, Benjamin began raising funds, assembling individuals to form the Board of Trustees, and was in charge of the construction. In 1751, the very first school of its kind was opened it's doors to boys that showed the potential to excel at school, regardless of how much money they had. The school was referred to as"the Philadelphia Academy, and Charitable School and enrollment grew rapidly. The school changed its name to as the University of Pennsylvania in 1779 as the Pennsylvania assembly took over control of the institution.

In addition to establishing a school, Benjamin Franklin also helped build an hospital located in Philadelphia. Although this wasn't originally Benjamin's plan, once the Dr. Thomas Bond, who was struggling finding funding and support to build the hospital, approached Franklin, Benjamin threw himself in the position. In order to get support and funds, Benjamin began to hold gatherings, reported on the situation in his journal as well as petitioned congress to fund the hospital. Through the support of Benjamin Franklin, the Pennsylvania Hospital that is currently operating, was established in 1751. The first patient of the hospital was admitted in 1752, however construction did not begin until almost three years later. Due to the compassionate treatment of the patients, not only offered treatment to physically sick but also to mentally ill patients and the number of patients grew exponentially. The hospital soon had many more patients than beds. In 1790, by the time Benjamin Franklin passed away in 1790 there were two wings in addition to more doctors.

VIIBENJAMIN FRANKLIN AND THE AMERICAN REVOLUTION

AN ENEMY OF BENJAMIN

Although Benjamin Franklin is known as a scientist, for most people, he's most well-known as the Founding father of the United States of America.

Benjamin Franklin was generally known for being a kind and generous gentleman. He was always trying to be a good person around the world and to be the very best human being he could be. There were however some individuals Benjamin Franklin would grow to dislike, and one of those people was George III, the King. George III.

When Benjamin was a supporter of the King George III after his reign was completed In 1760 Benjamin was beginning to believe that George didn't really care about his subjects. Benjamin was of the opinion that George III George III especially did not think about the well-being of the citizens within the American Colonies. In the time 1776 came around,

Benjamin Franklin wanted nothing to be associated with the King George III.

NO TAXATION WITHOUT REPRESENTATION

Although there were several steps that contributed to many of the steps that led to American Revolution, one of major ones is that of the Stamp Act of 1765. As a way to raise money in 1765, the British Parliament passed the Stamp Act to attempt to raise funds to get rid of financial debt. The act stipulated that a lot of documents made of paper, including pamphlets and newspapers, as well as wills and legal documents were taxed in a specific manner. Additionally it was stated that the tax can only be paid with the form of silver or gold coins.

The Franklin family was in London. Benjamin being in London during the time when his Stamp Act made its way through the colonies, it was believed that a rumor began to circulate that Benjamin as well as the Franklin family were a supporter of the Stamp Act. By way of a letter written by his wife Deborah, Benjamin learned many colonists formed a mob right on the front of the Franklin home.

It could also be an example of how Benjamin had at time time believed that the English alphabet was a possibility to improve because women at the time did not spell as well as Deborah.

A mob that was forming outside the house of his family perturbed Benjamin since he was concerned about the security for his loved ones. He was also concerned because Benjamin knew that these weren't those of the American Colonists he had ones that lived close and who he knew. In order to explain the shift in behavior that was not typical of those of the American Colonists, Benjamin began to blame on the Stamp Act.

Due to the American Colonists reaction to the Stamp Act, British Parliament requested a small number of colonists to explain including Benjamin Franklin, why the American Colonists felt the way they did on their reaction to the Stamp Act. In his address to the British Parliament Benjamin was asked questions and responded regarding the issue which he practiced previously.

A lot of historians believe that while Benjamin Franklin was speaking to the British Parliament over The Stamp Act, he was leading the way to believe the American Colonists were leading towards an independence from Great Britain due to some aspects he mentioned like that American Colonists would only pay the Stamp Act if they had to because of the loss of an conflict.

The second and final question Benjamin Franklin discussed when he was in with the British parliament was "What was the temper of American towards Great Britain before the year 1763?" Benjamin's reply to the question was "The greatest of all. They voluntarily complied with the authority of the Crown...They were not just respectful however, they also had a love for Great Britain, for its regulations, its customs and customs, as well as an appreciation for the fashions of its time. ..." The final inquiry Benjamin addressed to the Parliament the Parliament was "And how was their temper now," and the response to that question was "Oh, very much altered."

One participant in the British Parliament, the British Parliament was awash with attention as Benjamin Franklin was asking and answering questions himself during the discussion. Nowadays, many people are wondering that, given Benjamin's usage of the word "their" if he was suggesting in the direction of they were indicating that the American Revolution was approaching. But, there is no proof to be used to support this conclusion.

Benjamin wrote his letter after having been informed. The majority of American Colonists were also happy and also gave Benjamin Franklin a lot of credit for bringing about his Stamp Act repealed. When the news began reaching those in the American Colonies, they began to show their support on Benjamin and applauded the work he had done. But, a lot of the bitterness of those American Colonists towards Great Britain persist for a long time to come.

To the opposite coast the British weren't as excited regarding the repeal of Stamp Act. While the American Colonists were cheering on Benjamin Franklin, the British newspaper were releasing cartoons that put Benjamin Franklin in a bad image. They were enraged over the fact that there was a change in the Stamp Act had been repealed and maintained it was the case that American Colonists should pay for the British government's expenses for printing the stamps which weren't worthless. If Benjamin Franklin heard about this and was dissatisfied with their fury at the American Colonists. As with many American Colonists, Benjamin Franklin didn't believe these stamps that were now useless and the resulting loss to the British government to spend so much to lose, came from the hand of colonies. As a response, Benjamin published a strange and slightly unsavory story in the London paper.

A REVOLUTIONARY SONG

Because of the way that it was felt that the British were still feeling at times towards the American Colonies following they had resigned the Stamp Act was repealed, colonies started to be angry towards the British as their Mother Country. This didn't help when Great Britain continued to try to pass laws against the colonies, with their permission.

In 1767 In 1767, The Townshend Acts were passed on the American Colonies by the British Parliament. Parliament was of the opinion they were American Colonies were only there due to the fact that Great Britain supported them, especially in terms of finances. So, they continued to seek to increase taxes on specific goods that were produced in the colonies. They also tried to tax certain goods. Townshend Acts were a series of laws that imposed tax on certain materials including tea, papers and glass. As a result, American Colonists began to boycott those items taxed. Benjamin Franklin approved of the decision to boycott as he felt Britain was not a good country toward the colonies, and he wanted that attitude to be changed.

In the wake of the Townshend Acts, the American Colonists were still protesting in protest against the tax increases and other actions Great Britain was sending their to their. The violence in the American Colonies became so violent they British Parliament was forced to dispatch British troops to the colonies in order in order to end the riots. But, some American Colonists began to lash out at the British soldiers, insulting them and taunting the British soldiers. This was the moment when Benjamin Franklin began to warn the British Parliament to beware that they were pushing American colonists too far, and were making an unwise step by of sending troops into the colonies.

In 1770 after all the criticism that the American Colonists were giving the British

Parliament, it was ultimately decided to eliminate every single Townshend Act. One of the acts that was left was the tea tax. This was the time when the Colonists as well as British soldiers met in Boston. In March 5, this year, British soldiers started shooting at the non-armed American Colonists, at least the ones that came to throw snowballs at British soldiers. Benjamin Franklin was heartbroken and angered when he saw the details of what British soldiers did to American Colonists at the time in London. Franklin said that the soldiers were cowards because of their behavior.

In 1773 in 1773, in 1773, the Tea Act was passed on the colonies by the British Parliament. It led to an event known as the Boston Tea Party. This was the time when Benjamin Franklin realized that there was likely to be no reconciliation with the American Colonies as well as the country they called their Mother Country.

Also, it was during his time when Benjamin Franklin put a song in response to the sentiments of the American Colonists were feeling toward Great Britain. The song was entitled "We have an Old Mother" and contained the lyrics below, and were sung to the beat to "Which Nobody Can Deny:"

Our family has an elderly mother who

Peevish can be mature.

We are treated like kids who

It is rare to walk by yourself.

The girl forgets that we're grown-up and

Have a semblance of our individual.

Which nobody can deny, deny.

This is a fact that cannot be denied.

If we do not obey the orders,

Whatever the situation

She is angry and frowns,

And she is unable to keep her cool,

Sometimes she even hits us

A head-smack.

Which nobody can deny, deny.

This is a fact that cannot be denied.

Her odd order is,

It is not uncommon to suspect

The aging process has weakened

her lucid intellect.

Yet, she's an old woman

ought to be respected.

Which nobody can deny, deny.

That is something that can't be denied.

We'll be with her in court

to confuse all of them

Who is it, who can gain what she's got,

are often her enemies;

However, we are sure it can all be

We are our own. is gone.

Which nobody can deny, deny.

That is something that can't be denied.

Following the Boston Tea Party, the British Parliament adopted the Coercive Acts, which the American Colonists began to call the Intolerable Acts. As the British Parliament believed that the American Colonists required to be more in control and controlled, they sent additional British troops to colonies. Additionally they also shut off Boston Harbor. Boston Harbor and suspended town gatherings until American Colonists repaid the tax for tea.

In the midst of Benjamin Franklin still in London at this time and he is hearing what British citizens have to say regarding the American Colonists which makes him angry. "I have heard the government condemn Americans," Benjamin Franklin stated, "as the lowest of humanity, and as an alien species to those British from England. We are slammed with disdain and contempt...in my view, government ought to take care of all obligations imposed by the Armed forces."

In this moment, Benjamin Franklin felt and realized that war was likely to become the next step in the lives of American Colonists and he sought to prevent the situation before a conflict began. So, he asked to speak with members of the British Parliament, only to be refused. The Privy Council, which was an elite group of nobles that advised the King George III, asked to attend to Benjamin Franklin in a meeting.

The first meeting with Benjamin Franklin and the Privy Council took place at the start of 1774. In the course of this meeting, members of the council vented all their anger at the American Colonists against Benjamin Franklin, completely humiliating Franklin in front of a large number of British people. It is believed that Benjamin Franklin stood there like an unmoving rock all time and showed his displeasure towards Great Britain show after the session ended. He was walking with the chief council member that rebuked him at the discussion.

This was the moment Benjamin Franklin knew the war was about to begin since it couldn't be ignored or avoided.

BENJAMIN'S VARIOUS REVOLUTIONARY TASKS

In 1775 Benjamin returned to in 1775, Benjamin returned to the American Colonies, which was in the midst of the conflict. In less than 24 hours of having returned to American territory, Benjamin Franklin was appointed to the Second Continental Congress. When Ben began to attend the sessions of the congress, he recognized that he had been among of the very few who understood they American Colonies were required to defend their independence. This is why Benjamin was determined to do everything he could to support the cause of independence. This would frequently take him away from his work when they were not in his chair during Congress Continental Congress that the man slept very little at all, only to appear the time he was sitting at his desk. In one instance, the John Adams, the Founding Father and Second President of the United States of America, John Adams said that Franklin was frequently "fast asleep in his chair" in Continental Congress debates.

The duties Benjamin performed to support the Revolutionary cause varied. In the span of nearly two years Benjamin was able to raise funds and arms for the army. He also organized militias, organized the Pennsylvania militia, inspected troop numbers, developed plans for the defense of Philadelphia's navy as well as a brand modern mailing system. The British government was still in charge of the old system of mailing and therefore, nothing was given to American Colonists. So, Benjamin set up a network of cities in the colonies.

Additionally, Benjamin was also part of the committee that had in charge of drafting the Declaration of Independence. The task of drafting the Declaration of Independence went to Thomas Jefferson, who became the third president of the

United States of America, Benjamin helped with the writing process. As Benjamin Franklin was getting over an illness which resulted in a rash that was painful, Thomas Jefferson asked Benjamin Franklin to review the completed version of the Declaration of Independence. Benjamin generally did minor edits like straightening the words. But there was a major change Benjamin Franklin made, which changed the Thomas Jefferson's "sacred and undeniable" to "self-evident."

As the Second Continental Congress read and revised the draft of the Declaration of Independence, they weren't as friendly to the draft as Benjamin Franklin was. Through the entire process, Thomas Jefferson sat there watching and listening to the members from the Second Continental Congress tear apart the Declaration of Independence. Benjamin Franklin could tell that it was not just humiliating Thomas Jefferson but also saw that he was angry with the way in which he was edited. As a way to get Thomas Jefferson feel better about his situation, Benjamin Franklin told Thomas Jefferson an account. Since then, Thomas Jefferson and Benjamin Franklin kept in touch. Thomas Jefferson had high respect for Benjamin Franklin, especially for his assistance Franklin gave Jefferson when he was writing the Declaration of Independence. Jefferson was also able to have a warm regard in his heart for Benjamin Franklin because of the kind gestures Franklin displayed to Jefferson throughout his fellow members of the Second Continental Congress was tearing away his draft of the Declaration of Independence. It was true that Thomas Jefferson was one of the very few people to meet Benjamin Franklin before Franklin's passing. The get-together was held in the month of March 1790.

Jefferson visited Franklin at his residence The two of them had a conversation about their experiences as well as Jefferson said He was quite interested to know how Benjamin Franklin had been writing his autobiography.

The Second Continental Congress adopted the Declaration of Independence on June 2nd, 1776. the sixty-six Founding Fathers had signed the document on July 4, 1776.

After the signature, Benjamin Franklin reportedly stated, "we must hang together, or most assuredly, we shall all hang separately."

Benjamin as well as of the of the other Founding Fathers, was aware that they were guilty of treason that carried a penalty that included death. The same day, July 4, 1776 that the Declaration of Independence was read before the general public in the Pennsylvania's

Statehouse Yard. In 1776, the Declaration of Independence was printed and distributed to the colonists in as many colonies as they could. Colonists began taking down the statues of the King George III as they celebrated with joy. The celebrations for Benjamin Franklin, declaring independence from the Mother Country meant that everything had to be completed simultaneously time. Everything that was possible had to be done as fast as it was feasible. Like many Americans realized, and just as each Founding Father was aware of about the new United States of America would require victory in the Revolutionary War, which had been going on for less than an entire year, if they truly wanted to achieve their independence.

Chapter 5: Benjamin Franklin's Last Chapter

BENJAMIN FRANKLIN IS SENT TO FRANCE

At the time of the 1776 end at the end of 1776, when at the end of 1776, with Revolutionary War still raging on, Benjamin Franklin said goodbye to Philadelphia. As he grew older Franklin was required to travelling through France to garner supporters in support of his American cause. The time, Benjamin felt he was never going back to America and so the trip was a solemn goodbye to his loved ones, family as well as the newly-formed United States of America. Benjamin was aware that it wasn't the easiest task to try to convince France to join with America. United States of America, however, he was confident that the possibility was there and Benjamin was determined to win the trust of France officials.

Through the course of his time living in France, Benjamin heard of every news story about the Revolutionary War. He heard about the successes achieved by American forces as well as battles that were lost by the American troops. Through this time, Benjamin tried not to let his losses detract him from his work in France. But, it was particularly difficult after Benjamin Franklin received the news that the British had taken over Philadelphia. The city the place where his family resided. The city was where Benjamin Franklin's daughter Sally as well as her husband and their children resided in. Benjamin was shocked and frightened about the news. When the news first was announced in November 1777 Benjamin stood at the highest point of trying to convince France to join America. United States of America. For France it only heightened their fears of being a part of it was a fact that United States of America could never be able to survive in the absence of Great Britain. For Benjamin Franklin, it just ignited a greater desire in him to finish his work in order to help ensure there was a chance that the United States of America would triumph in the war and ultimately be able to gain independence. At this time the goal was not just protecting his own life, nor the lives of the other Founding Fathers and their families, but also the lives of his entire family.

It was just one month following the announcement of the British soldiers had seized Philadelphia at the time that the French recognized they could benefit from it was the United States of America would be a great ally for their cause. It was during the month of December 1777 that word was spread among Benjamin and France about the fact that all of the British Army was a prisoner of conflict. In the subsequent year, on the 6th of February 1778, 1778 on the 6th of February, 1778, the Treaty of Alliance was signed. It was the first sign the moment that France has finally joined with to the United States of America in the fight for their freedom. The time, at the same time the King Louis XVI began to not just think of his United States of America as an independent state, but considered Benjamin Franklin as the country's ambassador. Thus, instead of going back in America, which was the plan of United States of American after the agreement was signed Benjamin Franklin remained in France in the role of United States of America's ambassador.

It wasn't until that the Treaty of Alliance was signed that France began to understand what was happening with the American Revolution and the soldiers engaged in it, at the time that Benjamin Franklin started to become involved in planning military operations with France. While many were considering the possibility of shipping French troops into America, Benjamin Franklin was considering sending some of them to England. Benjamin began working alongside John Paul Jones, who was a former American Navy Captain. First thing they achieved was getting Jones and his team a larger vessel to take them toward England. Benjamin later gave Jones the strict instruction to be on guard for the British prisoner. Benjamin informed Jones that his soldiers weren't allowed to harm anyone in the group.

When Jones as well as his crew were sailing towards England and encountered Serapis which was the British ship that was en route towards America. Jones along with his crew together with the British aboard the Serapis took on the Serapis in battle on the 3rd of September 1779.

Jones taking the fight to a close by getting the Serapis. After the announcement of the victory of the Serapis, both France as well as America were euphoric about the victory. This was not just an American victory, but also an victory that was achieved at sea in England.

After British General Charles Cornwallis surrendering the British army to George Washington in October of 1781, in the Battle of Yorktown, the American Revolutionary War was over with the declaration that America was finally gaining independence. A peace treaty was still to be reached between the United States of America and England. This is why Congress decided to select John Jay, John Adams, Benjamin Franklin, Henry Laurens as well as Temple Franklin, who was Benjamin's great-grandson as members of members of the American Peace Commission to negotiate the agreement.

The negotiations were to last for some time and were close to being over a couple of times. In the beginning, England offered America limited independence, however the Americans did not accept. They were looking for England to shoulder some part of the costs for the war. England declined. It was only at the close of 1783 when the peace treaty was agreed upon and then signed. England was willing to compensate the entire towns and houses that they destroyed, if it was agreed that United States of America would take over the business and houses they took from the loyalists. England was also adamant about the US of America's total independence and ensured peace with the newly created sovereign nation. Peace Treaty Peace Treaty was signed by John Adams, John Jay as well as Benjamin Franklin on behalf of the United States of America. From the perspective of England the document was made by David Harley, who was one of the members of Parliament.

BENJAMIN FRANKLIN'S LAST RETURN HOME

In July of 1785 Benjamin Franklin retired from his ambassadorship and chose to enjoy his last years in the company of his family members in Philadelphia. Benjamin arrived on American land in September 1785, to be greeted with a warm welcome. Benjamin was elected as the sixth president of the Supreme Executive Council of Pennsylvania in 1785, on the 18th of October. The position was held by Benjamin from October 31 to October 31, 1787. Through the remainder of his life Benjamin was in a state of decline, mostly due to sickness. Benjamin was often in bed and even in his time as president of the Supreme Executive Council of Pennsylvania. When he told his daughter Sally she was ready to leave for the eternal land following her pleas to stay Benjamin Franklin passed away on the 17th of April, 1790.

IX EPILOGUE

In all his years, Benjamin Franklin was a man who lived before his time. But he was also one who didn't give himself as much praise as was due to him. Despite the fame and wealth He never gave up trying to be a good person every day to be the most ideal human being he could become. This is just one of the legacy Benjamin Franklin has left behind. Today, Benjamin Franklin is considered as one of the most well-known Colonial Americans not only of his time but across the whole time of Colonial America. There is a belief it is believed that Benjamin Franklin is the real father of the United States of America. It is due to the fact that Benjamin Franklin was the sole among those Founding Fathers who sighed all four of the documents to ensure it was possible that they could be signed by the United States of America could be truly independent from its Mother Country. The documents included: The Declaration of Independence The Declaration of Independence, The Treaty of Alliance, the Treaty of Alliance, the Treaty of Paris, and the Constitution of the United States.

Benjamin Franklin has many strengths that made him an eminent persona throughout American his time. One of them included determination, intelligence as well as compassion, understanding and a lot of hard work. Most of the time, Benjamin would mix these strengths to achieve his aim of being the most successful person he could become. In particular, his ability to negotiate not just only two peace agreements could have been achieved with the help of perseverance, intelligence and insight at the very least. We can today learn about the strengths of Benjamin

Franklin in a variety of ways. One of the biggest ways to gain knowledge from Franklin is to simply ask yourself the question Franklin asked himself each day Which good thing can I be doing now? Much like many other countries around the world, Benjamin Franklin lived in an ever-changing and changing world. He was not the only one to have lived throughout his time in the American Revolution, which is one of America's most bloody wars but was also an important component of both the American Revolution and the fight to gain independence. Benjamin Franklin could use every strength he had to build a better future not only for his family as well as for the majority of people living in America. United States of America. According to what Benjamin believed in his own daily life, everyone should be given respect and love. People deserved to be respected. This is especially true to this day. Every person deserves to be given reverence, respect and kindness.

Like all humans, Benjamin Franklin is just the same: a human. So, despite all of his strengths, he is not without weaknesses. Being a man of his time among his shortcomings was the fact that he didn't extend compassion and understanding towards all people. Although he believed that women should not be hurt, he considered that women were incapable to receive the same level of education like a male. In addition, throughout his time, Benjamin Franklin supported the institution of slavery but didn't see African Americans as equals. Although, at the time of his death of his life Benjamin did make a petition to Congress for the end of slavery but yet he didn't truly believed in equality. One of the weaknesses in Benjamin Franklin was forgiveness and anger. With regard to his son who was the oldest,

William choosing to support England in the American Revolution, Benjamin Franklin didn't forgive his son. Although he loved William, the friendship did not get restored following it was over. American Revolution because Benjamin Franklin was unable to release the anger that he felt against William in order to forgiving him Learn from the mistakes of Franklin through letting the anger go and accept the forgiveness of those who injured us in any way within our lives.

Chapter 6: Early Life Of Benjamin Franklin

Looking back at the earlier biography that of Benjamin Franklin we can go back to the beginning of the history of the Franklin family in the year 1682, a man known as Josiah Franklin along with his wife, left the country of his birth Northampton county, England and moved towards Boston. Josiah Franklin was the father of the great Benjamin Franklin. His wife died shortly following their move to Boston. That left Josiah and his seven children in utter isolation. In response to the need of the children Josiah Franklin remarried a lady, Abiah Folger, who was a well-known colonial woman.

The time of the birth of Benjamin Franklin, it was the 17th January 1706 that Franklin was born in Boston, Massachusetts. In that time his father worked as an tallow chandler and the soap boiler maker, and became famous for his role for being an

Immigrant from America who came to England. Benjamin Franklin was born from the second marriage of Josiah Franklin so Abiah Folger was the mother of 10 children. Benjamin was the eighth child within the family.

A child who was ill-treated and in a crowd

It is a fact that in a family, with 13 children, the prospect of enjoying luxuries was nearly impossible. All children's lives was arranged in such a manner that even the essential needs were not met with ease. This meant that it was impossible to giving the kids a good education. Benjamin was a regular student just for a couple of years. Benjamin's family couldn't pay for the tuition of so many children, and like other children, Benjamin also left school in the early hours. At the age of ten, he was taken to his father's store to assist his father.

Though Benjamin Franklin knew about his poor family and the lack of parents who could provide their family with the necessities of living, the man was unsatisfied and agitated within the store. In addition, he did not want to be by his father's work however he was also averse to the soap-making business. He became very angry whenever his father would bring the boy to several establishments in Boston. He was told by his father that he should watch and observe craftsmen in action so that they could develop an interest in soapmaking and also master the fundamentals to assist his father.

There was an urge to learn in Benjamin's brain from the time he was a child, and the absence of education caused him to be agitated, however Benjamin was aware that he couldn't not pursue a formal education. The drive to education led him to increase his informal education. He would have to be spending all day in the shop helping his father, however whenever there was time to read, he would start reading what could be found. Once he realized that education in formal schools was not his capabilities, he began to search to learn in a non-formal way.

Under the guidance of the elder brother at the Printing shop.

The first years of his life were filled with struggle to make a living, so after two years of working at the soap store that his father owned, Benjamin was then sent to the printing business that was run by James and his brother. Benjamin was taken to be an apprentice. Then his brother became allowed to establish the newspaper. Within the newspaper Benjamin began to write under a pseudo name of Mrs. Dogood. The brother did not know that the letters sent to their newsrooms were written from Benjamin

Franklin's Brother Benjamin Franklin. With his incisive style of writing, he proclaimed his unwavering support the freedom of speech. He was forced to go the pseudonym due to the fact that when he was first thinking about creating something on newspaper, his brother scolded him severely. The time the public thought of him as a stupid child who was able to follow orders given by his father. Yet Benjamin Franklin was quite desperate to create something thought-provoking to stimulate. He might have been aware of his extraordinary talents that was a gift from the nature of his work. He stumbled upon this method and was able to send his works under the name of widower. The character was sketched with great care and continued to write under the name for a long time.

The escape

While Benjamin Franklin discovered a method to deal with his own talent by writing for newspapers, Franklin was extremely unsatisfied with his brother because Benjamin was treated by Benjamin very cruelly. If he would get upset by a print press job, he was known to smack Benjamin brutally.

Then, he needed escape from his master. In reality, when James discovered his brother Benjamin that was writing under the title Mrs. Dogood, he became in rage. He was unhappy with his brother as his writing was praised by viewers. Then, Benjamin was severely beaten by his brother and Benjamin was unhappy with the way he was treated by his brother, fled.

Franklin fled from home. The only thing he owned were his ratty clothes and wet clothes on his back. There were only three dollars in his wallet. He was in a miserable state as he entered an establishment and left with three loaves of bread. After arriving in New York he started looking for a job in the field of printing since his experience was in this. At the time there was just one printer in the town of William Bradford. He was unable to find work, he travelled into New Jersey, from where the move was made to Philadelphia through an excursion on a boat. He travelled to Philadelphia and landed at his destination at the Market Street wharf. It was the month of October in 1723 at the time he was just a teenager was scouting around Philadelphia to find food, employment and a home to call his own.

It was at the time leaving the home of your parents was considered illegal. This was the time of the early days in America where the social position held by one was believed to be the most crucial element of their existence. Therefore, those who fled were considered to be a failure and poor standing. When he finally arrived, the entire money that he had been able to spend was used in buying rolls. The man wandering around street corners of Philadelphia would one day be recognized as the president of our nation.

Chapter 7: The Voyages Of Benjamin Franklin

The first time Franklin traveled to Philadelphia was certainly risky and dangerous. It was remarkable to think that a person of an education so limited and no prior experience could flee from the city of his birth Boston and set off into a new and unfamiliar city in a single-handed manner. In this time the city was his choice, i.e. Philadelphia was believed as the most important city in America that was also a bustling center with regard to business and culture. At the time that Franklin arrived in Philadelphia, he was optimistic and confident of the prospects of succeeding. Numerous American writers have wrote about the confidence possessed by Franklin and have written about how anyone can fulfill all of his desires by focusing on them and embarking on an entirely fresh journey and forgetting all the mishaps. Franklin didn't allow the lack of resources be a hindrance for him, and he began working completely from beginning from scratch. Franklin's ability to be a master of learning and study helped him to excel in a unique way.

It was not easy to find work in a place that was not known for Benjamin Franklin, he made his way there and got the job of a student printer. Though his time together with James, his older brother James on the printing press was somewhat disturbing, when it came to the education of the printing process is concerned, the times taught him much about the printing processes.

Benjamin was very skilled when it came to his job. He excelled even at such an early age that the Governor of Pennsylvania said he would set his business up for himself in the event that young Franklin went to London to purchase printers and fonts.

Franklin was able to go to London however the governor did not keep his word and Benjamin had to work for some time in England working on printing.

Benjamin franklin at London:

The Governor of Pennsylvania was able to see Benjamin working with such skill, He encouraged him to put in place an own buying house. Benjamin was asked to travel to London to purchase all the equipment needed. He was told by the governor that upon his London trip, he'll assist Benjamin to set up the publishing company.

At the time of 1724, Franklin relocated to London following the direction of the Governor. He was instructed to purchase the necessary items from booksellers or stationers as well as printers. As soon as he arrived England the British embassy informed him that he was manipulated by the governor because his letters of introduction and financial statements were not delivered. This was an additional challenge for Franklin since he didn't meet anyone from the city. He recalled the time when he first moved to Philadelphia in the midst of a homeless situation and had to be dispersed.

In this time when he was in London the man didn't feel so depleted. The lessons of life had taught him therefore, even though he was required to locate a job within London but he also made sure to take all the pleasures of the London city. London. The bustling and vibrant city of London was a magnet for him. He went to theater shows as well as mingled with locals at various cafes and never stopped his love to read and explore. The self-taught swimming instructor who was clever enough to design the swimming flippers of his own made out of wood. He competed in long-distance swimming races on the Thames River.

In 1725, in 1725, that Franklin wrote a pamphlet entitled "A Dissertation upon Liberty and Necessity, Pleasure and Pain". This was his very first pamphlet with which Franklin claimed that human beings do not have the right to free choice which is why one can't hold their actions accountable. their choices.

While he was able to travel to London in deceitful motives, however, this garment proved as an expansion tool of his understanding about any kinds of printing. It was fortunate that the location it was that he stayed in, Little Britain, was thought to be the center for printers and booksellers in the time. This was not all, but it also was the main centre of discussion on politics and religion. The place where Franklin came across John Wesley, the founder and preacher of

Methodist congregation, who was a preacher in nearby. Franklin was able to reflect about religious issues when he listened to his sermons.

Franklin always came across as the ultimate fighter. However difficult his circumstances seemed at him, the man did his best to get through the day. When he was on in the streets of London He did not bear any burdens of the lack of resources that life gave the world. He saw every part of his existence as a learning opportunity and gained as much knowledge as possible. Then he got a job in London in Samuel Palmer's print shop. When he was at work, at the beginning of 1725, He worked in the printing shop to print an updated publication of Religion of Nature Delineated by William Wollaston. It was his responsibility and he was required to release the book, Benjamin Franklin was greatly unimpressed of the ideas formulated by Wollaston. He was not only unimpressed however, his own inner writer wanted him to spread the thoughts of his mind and to spread these ideas out to the outside world. Again, Franklin chose the route took him in his early growing years. He wrote an essay anonymously to help people comprehend his ideas that came out as a reaction to Wollaston writings. He titled his essay "A Dissertation on Liberty and Necessity, Pleasure and Pain". The essay managed attracted interest of the free thinkers as well as liberals in London.

In the latter time of the year Franklin was employed by a printing company in London. He then decided to collaborate together with Thomas Denham, who was an Quaker merchant.

Get in touch together with Thomas Denham:

In the summer of 1726 Franklin returned to America alongside Denham. Because Benjamin Franklin was an incredibly quick learner and had an intelligent mind, he mastered everything about Denham's enterprise quickly.

Thomas Denham was actually a merchant from Philadelphia that played a major contribution to the personal life that led to the life of Benjamin Franklin. Franklin mentions Denham as a close friend or father figure as well as a benefactor throughout his life. This was a huge help to Benjamin Franklin quite devotedly even after his return back to Philadelphia after 1726. In 1726, when Franklin returned in 1726 to Philadelphia, Denham opened a brand new business and hired Franklin as the bookkeeper and clerk. The two also resided in the same house.

Denham was the man who helped Franklin with his psychological and moral development to help him move smoothly from a sluggish youth to a mature, young adult. Denham helped Franklin recognize that the pursuit of success and virtue in this world isn't something that is easy to accomplish. In England where Franklin came into contact with Denham and he was Denham that showed his honesty and made him realise that the Governor Keith was lying to his trust and thereby guided Denham to England. Denham's initial generosity to Denham toward Franklin was further substantiated in the form of a lifetime commitment in the event that Franklin was able to stop the plot of Andrew Hamilton, who was an acquaintance of Denham and was a well-known attorney in Philadelphia. The time it was Denham who advised Franklin to find work as well as to gather money to enable him to return to America as did Denham. When he returned home to Philadelphia, Franklin worked and resided together with Denham until the passing of Denham that was long ago, however in 1727. At this time that Franklin started the printing business that was run by Denham. However, Franklin was devastated by the death suddenly of Denham. Within such a brief time, Franklin had been able to form a solid connection with Denham and they both enjoyed speaking and working.

Franklin was working on the printing presses of Denham, Franklin found a partner in business, who was identified as James Meredith. Both were keen to open a printing business with Meredith, and Meredith was able to get a money by his parents. They had plans to begin creating a newspaper. The town had a man who lived in the town Samuel Keimer who learned about the plan, and immediately printed his newspaper. Franklin was very irritated by his plan and composed a series of essays to an opposition paper. Franklin held so much influence in his speech that Keimer could not sustain it over a long period of time and in one year, he had become financially bankrupt. Then, Meredith and Franklin started the Pennsylvania Gazette and made it one of the top papers of the time. In 1730, they reached another important milestone of their accomplishment after they were awarded an agreement to print the official newspaper from the state. In the next few months, Franklin took over Meredith and, in very little time Franklin was one of the most well-known and successful printers in the town.

A meeting with Deborah Read

Benjamin Franklin first met Deborah Read just a few days after his first journey to Philadelphia the year 1723. In that time Benjamin Franklin was in the seventeenth year, while Deborah was a teenager in her sixteenth year. Benjamin Franklin while writing his autobiography has written lots about their meeting, romance and their wedding. Though Deborah's story gained attention to historians because of her connections to Benjamin Franklin, yet in 1723 when they first met Deborah surely enjoyed the benefit over Franklin. Her family was middle class. family where her father worked as a carpenter in the city. Due to his tireless work her father managed to build a house within Philadelphia which was a busy city. However, at the similar time, Benjamin Franklin was the unemployed, runaway trainee who had no possessions, and very little cash. Franklin was enjoying an evening meal along with his saved "puffy rolls" when Deborah first met him.

At first, she was snubbed and rejected from the families of Deborah Franklin went to England in 1724. The year was 1726. Franklin returned to Philadelphia where he rented at the home of a Godfrey family. He formed a relationship of love with a relative of Godfrey's relatives. After Franklin proposed a one hundred pounds dowry they resigned in hopes that they would leave and release the family from the obligation of the dowry. Then Franklin removed himself from the whole thing and gave the offer in favor of his Godfrey family. This was the moment when Franklin turned his attention toward his Read family. Though there were many challenges along the way, in 1730 Benjamin Franklin and Deborah declared their marriage official. Franklin writes in his autobiography of the couple that Deborah was a smart and trustworthy mate who worked with him a lot. She was in the shop with him and stood by him all the thick and thin.

Chapter 8: The Newspaperman Working At Pennsylvania Gazette

In the time that Samuel Keimer was heading it the paper, it was named "The Universal Instructor in All Arts and Sciences and Pennsylvania Gazette". However, Samuel was unable to stand the arduous demands of a the printing industry and after one year, he was forced into financial ruin. It was at this time that Benjamin Franklin took the charge of the paper. In that time it had just 90 subscribers.

Franklin's contributions to the Pennsylvania Gazette

When he took control of the paper, Franklin made many changes in the beginning; one of them was regarding what the name of the paper was that was changed in order to create The Pennsylvania Gazette. Then, he introduced a number of adjustments to the structure of the newspaper. At first, the newspaper had an area titled "Universal Instructor" which consisted from the week-long page that was part of "Chambers' Encyclopedia". After Franklin's rule, the paper became extremely profitable. The newspaper's name was changed later in the form of "The Saturday Evening Post". Franklin was certain that the newspaper would be an all-inclusive package that could be enjoyed by every kind of reader and cater to all kinds of interest. So, he made sure to include local news, a variety of excerpts from The "Spectator", verses, humorous attacks against rival newspapers and moral essays by Franklin as well as intricate hoaxes and political satire. Franklin was a lover of printing, and it became obvious within the next few years when Franklin earned his place in one of the top-rated and well-respected newspapers. The newspaper was a struggle for him to bring the newspaper to greater heights and was very accomplished.

In addition to printing, Franklin also continued with his penchant for writing. The tense little essays that Franklin composed in Pennsylvania Gazette had been able to hold an permanent place in the history of American literary. As journalistic part of the colonial period, they still have the historic position.

Franklin is a pioneer in the the publishing industry

Franklin began his career at a time that was not attracted by publishing job. The early years were spent with his older brother, however at the end of the day, his publication was his main source of fortune and fame.

Thanks to his massive efforts, he elevated the publishing business to highs of the industry. One of his greatest achievements is the launch of a German newspaper for the first time in the history of the United States. While his venture didn't succeed, it was regarded as one of the first actions in the publishing industry and laid the foundation for German publications to be published in the United States of America.

A major change created in the early years of Benjamin Franklin in the publishing business was the usage of cartoons with political themes. Franklin produced a cartoon about politics called Join or Die. The cartoon first appeared in 1754 by the Pennsylvania Gazette. This was the first time that this cartoon in the Gazette is thought to be the first illustration of colonial unification conceived by the British colonist of the United States of America. The first version of the cartoon was made of wood depicting a snake divided into eight pieces. Each piece was identified with initials that indicated the names of the American colony or region. The cartoon was released alongside the editorial by Franklin. The editorial addressed the subject of the disintegration of the colonies. By utilizing the editorial, Franklin was able to make his case for the importance of a united colony. No one knew this illustration could garner so much popularity until it became the most iconic symbol of liberty for the colonists during the Revolutionary War of America.

In the 1730s, the popularity and popularity of Franklin was growing quickly. The reason for this was his decision to launch the Poor Richard's Almanacs. The year 1732 was the year he began this publication. Franklin published it continuously for the next 25 years. It also was a complete set of information about astronomy and weather forecasts, poems Proverbs, witty maxims, and proverbs written by Franklin. The work he did earned him good money and there was the time where Franklin was regarded as one of the wealthiest men of Pennsylvania.

There is no one other than Franklin who was able to achieve the most in such an incredibly short time. From the days of a runaway who was homeless to being the wealthiest, he went through numerous challenges and ups, but during the course of his quest, he always satisfied. A major part of his character was to become self-sufficient and an ardent fighter. From the financial gain to popularity, he was obsessed that matched these traits.

Chapter 9: Benjamin Franklin As A Diplomat

Societal and Communal Services:

Benjamin Franklin was more than a remarkable person, and he extended his successes to Philadelphia Gazette and Poor Richard's Almanac. Franklin gained acclaim by the establishment of the his first firefighters in Philadelphia as well as improving the postal service and establishing an open-air hospital. These accomplishments are all tied to his popularity and local image, however it was the extraordinary set of discoveries and scientific research and the mapping of the great Gulf Stream which put him at the forefront of world forums. As the years passed by, time and his reputation grew, he was able to be trusted as a diplomat in the court and salons of Europe.

Franklin helped in the establishment that was part of the Pennsylvania militia, and was able to raise enough money to build an urban hospital. He also organized an active plan to create streets and set up lights for the streets of Philadelphia. In addition, it was due to the tireless efforts from Franklin and his associates that it was through his efforts that the Academy of Philadelphia was created. The institute was established in 1751 and later elevated to the status that of University of Pennsylvania in 1791.

Franklin is also acknowledged as a key figure in the post-colonial postal system. In 1737, during the British government Franklin was appointed the postmaster in Philadelphia. From 1737 to 1753, he was promoted to being the joint postmaster of every one of the American colonies. The position was achieved because of his determined effort to enhance the system in every way possible. He came up with ways that no one else had thought of in the past. He came up with a variety of measures that would expand the postal service. In 1774, his dismissal was imposed by the British government for believing that he was too compassionate and concerned about the colonial concerns. The year was 1775 and Franklin was named the postmaster general of the Continental Congress. The postmaster general was his for the first time in the historical history of America. It was an amazing achievement since under his title, it was his authority to take charge for all post offices between Massachusetts up to Georgia. His authority was his until 1776. Many historians and specialists have pointed out the significance of Franklin as an early pioneer in the field of postal services. Franklin's knowledge and leadership skills that enabled him to chart the foundation for the growth of our nation's progress within this area.

Benjamin Franklin and Union Fire Company

Following a visit during his visit to Boston, Franklin observed that those who lived in his birth city were more prepared to respond in the event of an emergency fire than those of his beloved city of Philadelphia. The thought stayed in his head and after the time came to return, he initiated discussions with the Junto group, an organization of goodwill dedicated to improving the quality of life and public services. Franklin discussed with them concerns he had and demanded ideas on ways to enhance the way they fight fires to guarantee safe fights in the event of flames.

Franklin was also involved in the public domain, and ran awareness campaigns in order to educate the people of the urgent need to develop firefighting techniques. Franklin also published a piece on the subject in the Pennsylvania Gazette and guided people on how to fight fire with a more professionally. He also pointed out that the most significant obstacle to fighting fire is the fear that comes with the fire, and this would have to be tackled by professionals assistance.

Firefighters who were Goodwill and Amateur at the time did not suffice but they were adequate. Franklin suggested that a club, or society was formed that included the men who were active and fitting to the Fire Engine; who will be charged with fighting flames in cities when it occurred.

He also suggested that chimney sweepers be certified by city officials and responsible for the entirety of their activities. He also noted that even though there were firefighters working in the city, they continued to work as they would any other work. Franklin advised that a drive to serve others was required to be cultivated in these firefighters to ensure that at time when they needed to, they can use their expertise with more determination and passion. Franklin pointed out that the firefighters should be trained frequently in order that, in the event an emergency, their response time can be reduced completely. To accomplish this the firefighter arranged for smaller meetings and discussion and then published his findings in his journal. Franklin believed that, before taking the action can be taken, one must get trained internally.

It was through the more deliberate effort from Franklin that a small group comprised of thirty members was formed together to form the Union Fire Company in December 1736. One of the main actions that Franklin suggested Franklin was to arm them with wooden buckets as well as stronger bags and baskets to ensure safe packing as well as efficient transport of merchandise. The men were called blaze fighters who met every month to discuss the best firefighting and preventive techniques. The homeowners were instructed to have the leather buckets inside their homes to ensure that everyone could respond to an emergencies.

As they observed the development of this group, others were urged to join the organization, but were instructed to start small businesses on themselves so that an extensive network of security could be put in place to safeguard the entire city.

In the nick of time in the past, Philadelphia loved the establishment of a variety of fire organizations like the Heart-in-Hand as well as the Fellowship as well as the Britannia. It was the extraordinary intelligence and the management of

Franklin who established Philadelphia among the most secure cities in the world.

Start of the diplomatic trip:

The year was 1757 and Franklin began his diplomatic career. The time that he was appointed to the Pennsylvania colonial assembly to go to London to represent the colony to resolve a tax controversy. He was tasked with settling the issue with officers from George II. George II. The time he was away from the King's palace lasted to 1775. After his arrival in America and was elected the representative of Pennsylvania at the Second Continental Congress. He was a key player in the creation of a draft document to draft the Declaration of Independence. Due to his worldwide repute his name was chosen by France at the time of the Continental for the role of an official to the Colonies. This was during the time when American Revolution was on its course, so he had an important role during the critical time. In 1778, he was elevated to the rank as Minister Plenipotentiary. After presenting his qualifications before the French court, he was the first American Minister, which was equivalent to the status of an ambassador today.

Franklin was blessed with the natural ability to persuade and negotiations, which was used to its fullest to make France be a signatory to an Treaty of Alliance with America. The Treaty of Alliance was the biggest sign of approval for the newly formed republic. The treaty also allowed the country to appreciate military, financial and even political support that helped the colonies later in fighting off the British in the independence War. Franklin also served as Minister Plenipotentiary to Sweden and negotiated together with the Swedish government on an agreement that was concluded in 1783.

The treaty was named Treaty of Paris, Franklin was the main participant to represent his country, the United States, which eventually separated the colonial bonds with England.

In 1785, Franklin returned in 1785 to America in 1785, where he was active with societal activities for Pennsylvania. Franklin was a prominent participant of his participation in the Constitutional Convention of 1787 and was a witness to the sanctioning of the newly established Federal Republic in 1790.

Franklin as an American Diplomat:

Franklin received a post as ambassador to France in 1776 until 1778. In this time Franklin was tasked with the crucial and difficult job of winning French support for American independence. French elites and intellectuals regard Franklin to be the most accurate representation in the New

World Insight. The man was an ambassador for the French. French ambassador, who earned popularity throughout America as well as France. His love of Franklin is evident in medallions, ring or watches as well as various Snuffboxes. In fact, Franklin styled fur caps were in French style for a time time. The acceptance of his diplomatic skills became a challenge after his his first American victory in battle at Saratoga. The battle helped convince France to recognize and even support American independence, which culminated with the French partnership with the thirteen states. Franklin is believed to set the standard to all American diplomats. His expertise and diplomacy earned him many successes as well as a reputation in America. American nation. The nation's reputation was elevated at this crucial time was the result of Franklin who will be remembered by the American country will never be able to forget.

Articles of Confederation:

The Articles of Confederation assisted as an official document which recognized the role and importance of the nation's government in America following its declaration of independence of Great Britain. The document recognized a central authority that, though not entirely, restricted each state from displaying their distinct foreign diplomatic diplomacy.

Benjamin Franklin had proposed a proposal to create Articles of Confederation with the Perpetual Union. While a majority of individuals, such as Thomas Jefferson greatly supported this idea of Franklin However, there was many opposed. Franklin presented his proposal in before Congress on the 21st of July, however he was adamant and stated the plan be considered as an informal and preliminary proposal to be developed into more formal proposals when the time the Congress might be attracted by the idea. Congress made the proposal available as a subject for debate.

Following the Declaration of Independence following the declaration of independence, it became apparent for all members of the Continental Congress that a formal established nation's government was essential in order to conduct all State affairs smoothly. Therefore, discussions on the structure of the government began with Congress on the 22nd of July in a dispute over a variety of issues of importance, like the voting proportionality between the different states. Numerous different issues were identified that became the source of divergences. These divergences delayed the conclusion of talks for confederation until October 1777.

In that time, the capture of Philadelphia by the British was a factor that made the subject important. In the end, the representative came up with the main Articles of Confederation, according to that an agreement was reached regarding proportional tax burdens, as well as states-by-state elections. But the question the use of western land by different states had not been resolved. These documents were given to the states for their adoption and amending. While a few groups protested against the issues that were discussed in the articles, everyone agreed that at this point the creation of a federal government was required.

The Treaty of Paris:

Benjamin Franklin vetoed casual peace suggestions made by Great Britain for an agreement that would give thirteen states that had a low degrees of autonomy, while belonging to the British Empire. The year was 1782. Franklin was a firm believer in American independence, and he was against any attempt the idea of a peace treaty independent of France as an unwavering ally of America. Yet, Franklin was in favor of talks to bring the war to an end with British powerhouses. It was the best wisdom of his because conflict eventually destroyed the generations on both the sides. But, he was also in favor of a high-profile job to America. Franklin was along with two peace commissioners, John Jay and John Adams and John Adams in 1782. The two men then conducted informal talks with British authorities, and dealt with the issue with tact.

Even though Franklin required the end of Canada to create independence for America He was also aware that the British government under Lord Shelburne did not have the capacity to accept the deal. Negotiations were conducted over a period of two months that consisted of a number of difficult bargaining sessions. The end result was the first peace treaty. In this treaty, British officials accepted American independence as well as the legal boundaries, recognized the thorny concerns regarding the fishing rights along the shores of Newfoundland and dealt with prewar loans of British creditors. They also promised payment of the property that was lost in the war, and agreed to the removal of British troops from all thirteen states. This initial article was signed on the table in November 1782. However, they were deemed valid after France and Britain agreed to a similar treaty. French Foreign Minister Olivier de la Roche quickly reached a deal to sign the initial articles of agreement along with Great Britain in January 1783. A formal peace treaty was concluded in Paris on September 17, 1783.

Chapter 10: The Works And Achievement Of Benjamin Franklin

Benjamin Franklin was of the conviction that the main purpose of scientific research and the resulting developments is to help ease the daily life of a common man as it is possible. So, he always focused for the most effective methods of communicating how science can benefit people of all ages. He was a firm advocate of the idea that our the natural world operates on the principle of simplicity to the human race, scientists are obligated to find that ease using various innovations and discoveries. Franklin loved studies in science from the time of his birth however his busy agenda did not allow him to take it up to its fullest until he quit printing in 1748.

In the time that Franklin published the Poor Richard's Almanac in the early years, he included detailed information about the weather. Franklin was interested in the subject of ocean currents that caused him to draw on the current Gulf stream. The reason for his research was the time spent by ships on different routes in delivering mail to England as well as other parts of America. While it was part of the information available to whalers as well as explorationists, Benjamin Franklin published it in marine observations to be accessible to the public at large.

Franklin was always driven to study science in the interest of improving public wellbeing. He was never contemplating having any sort of patent right for his ideas and discoveries.

Learn more about the power grid

In the summer of 1743, Franklin went to Boston where he was born. Being a knowledgeable researcher since childhood, the show he attended was the science exhibit there. On the show, the show featured the Dr. Archibald Spencer, who originated from Scotland and was the subject of many science-related phenomenon. While it covered a variety of sciences, it was the electric portion of the show captivated Franklin most. This consisted of the demonstrations of static electric currents.

The audience was amused by the demonstration presented by Professor Dr. Archibald Spencer but Franklin thought there was something lacking or wrong with the demonstration of static electricity. Franklin was swayed by a desire to know the reality, and so in 1747 Franklin purchased a large glass tube to effectively generate static electricity. It was purchased from Peter Collinson. After studying the phenomenon then he outlined it using a simple language so everyone who was interested in this field could grasp the static electricity. The current was shown as a kind of flow of fluid, whose flow is in the direction of an upward to downward direction. body. He also invented the concept of the concept of "battery current" and demonstrated the properties of negative and positive currents.

following the sudden discovery made by Franklin after his rapid discoveries, electricity was a important field of study. Many scientists began taking keen interest in this field. As a result that the electricity field saw a significant increase in progress when compared with the previous.

Franklin was of the opinion that electricity, as the most fundamental type of natural energy requires a lot of research. When he was 1753, Franklin was at 47 years of the age of his birth, his revolutionary efforts in the realm of science were appreciated by the British Royal Society, who honored

Franklin's work in electrical engineering by awarding him the Copley Medal, which was the most prestigious award at this time.

Lightning studies

In 1751, following a string of efforts of continuous investigation and testing, Franklin could publish the result of his work as an article he described"Experiments and Observations on Electricity "Experiments and Observations on Electricity". The book was read by thousands of viewers across Europe as well as Britain and the British Isles, who said it was the most important texts that helped them comprehend the fundamentals of electric power.

Then, in 1752, Franklin could carry his most famous scientific breakthrough that became an innovator in the scientific world. In the wake of this discovery

Franklin believed that natural lightning is an electrical phenomenon. Franklin knew the possibility that people might be interested in the proof that reason, which was the basis for his study. The experiment was planned in the light of his previous findings. It was based on the concept of static charge as well as the contrast between the object that is blunt as well as a pointed object that carried the charge. He devised a set-ups that he described"a lightning rod. In the present, many electrical experiments rely on Franklin's lightning rod in many designs and configurations. The setup was not used for patents by Franklin since he believed his research should be beneficial to everyone who was interested, without expense.

Today's researchers and scientists claim the basic knowledge of Franklin on lightning was the catalyst that led to scientists progress in this field. Because of this that lightningproof and safe structures have been built across the globe. As a result of his research findings Franklin certainly was an inspiration to all.

Refrigeration studies

A different area of research conducted by Franklin was the principle of refrigeration. In 1758, when Franklin was working alongside John Hadley in England, He studied the concept of refrigeration in the condition of evaporation. In addition, he designed a full setup to carry out an experiments on the entire concept. Franklin through constant experimentation and investigation, proved the rationale that explains the phenomenon of refrigeration. Franklin demonstrated that in any liquid there exists a range of molecules that differ according to the range of energy as well as the molecular sets. Certain molecules have higher energy capacity, while others have lower. The molecules with the most in energy capacity will first leave the liquid's surface and therefore evaporate very quickly. This means that the ones that stay in the liquid are ones with a less energy and, consequently, temperatures drop overall. The ideas and the demonstrations of Franklin are nothing less than an innovation in the field of physics as well as science. The basic concepts are thought of as the fundamental basis for refrigeration and molecular research.

His most important contribution to science was his ability to think about it as a method to extend services to humanity. He was not a fan using commercially-oriented sciences, so it is impossible to imagine Franklin running a profitable businesses in the field of copyrights and patents. Everything he did was made available to the public as open source that could be utilized and further improved.

Meteorology

The year 1763 was the time that Benjamin Franklin participated actively in discussions conducted by colonial scholars on the different effects of deforestation on the local climate. The forests of the past were cleared extensively in order to allow farming to colonies. American colonies. Franklin pointed out that area cleared of the forest is more prone to absorb heat. As a because of the temperature rise, the snow on the mountain melts faster. The majority of other scientists agreed with the same conclusion. However, Franklin was of the opinion that, before deciding an ultimate judgment, a lot of analysis is required so that an accurate judgment is rendered. This is why Franklin suggested a new field for discussion within the study of climate related the deforestation.

Alongside the meteorological skill, Benjamin Franklin also made the first scientific chart of the Gulf stream in the North Atlantic. Franklin proposed that trade winds create the formation of the Gulf Stream by pushing warm waters into and into the Gulf of Mexico, from that point, they leave the route through the Florida Strait and continue to create an Gulf Stream. The year 1775 was the time Franklin was poised to make his route to England and he stepped out of as a thermometer within the Atlantic and found that the Gulf Stream was 6deg F higher than the other parts of the ocean. It was this finding that prompted Franklin to create the groundbreaking current chart it was certain to become to be a milestone in the history of meteorological research.

In the final years of his life Franklin was engaged in research in connection with the potential impacts of eruptions of volcanic origin. Franklin believed that volcanic eruptions could have the cloud structure in a different position as well as weather patterns and the electrification of clouds. He researched the history of a volcano eruption throughout the globe. Franklin further suggested that there is a reduction in the amount of solar energy absorbed from the Earth's surface. Earth in the aftermath of every eruption due to the emergence of particles that follow the eruption. Including Ash. The particles then become obstructions between the earth's atmosphere and the sun, and the amount of energy at the point of no return is reduced.

The renowned Bifocal spectaclesFranklin himself was one who needed glasses for the majority of his time. While wearing these glasses there was always an uneasy feeling since the lens that he used to read by his eyes was always full blurriness as he gazed upwards. Because his primary job was printing, this was very frustrating.

Always irritated by this issue He came across the solution at the age of 1739, when he was 33 years old. He developed bifocal lenses that had split lenses. The lenses in this instance lens would be able to have two different distances of focus. Therefore, if a person were to be scanning through the lower part of the lens it could help in reading. In contrast should he look through the upper part of the lens it could be useful in observing long distances. It was an innovative innovation in the field of optics, solving the issue that so many others face. The result was of constant thinking and analysis which led to the fact that Franklin was able to discover solutions in the most effective manner.

The year 1743 was the year that due to the extravagant interest and work by Franklin The American Philosophical Society was formed. The term "philosopher" was employed for scientists. The Society was established with an goal to expand various options and facilities as well as a scientific forum in order to try out the new concepts. Franklin was also a tester of electric theories.

The Size of the Units of Matter

Franklin conducted a remarkable experiment using surfactants on a pond in Clapham Common. It was accomplished with the help of a tiny quantity of oleic acids, an natural surfactant, and allows the development of a strong film on the surface of the water-air contact. After that, he carried out every measurement to figure out the quantity required for covering the entire lake. After determining the total surface, he was able to determine the size of the film that was formed through Oleic acid. He then made suggestions to define the elements of matter, in order to establish the precise size and type of matter.

The Revolutionary Franklin Stove:

His ever-growing curiosity regarding science resulted in advances in many areas of research. The same was true for the issue of heat transfer, which caught his interest. He analyzed the layout of the typical stove and concluded that the design is inefficient in a variety of ways. The primary reason that he identified to explain this was the loss of heat that exceeded the amount required. He was always exploring the realms of exchange and heat recuperation therefore he decided to redesign the entire stove

The design of his stove on the notion that, when hot gases ascend along the flue, exchanges of heat will occur between hot gases as well as the cool air within the room. In the end, heat travels from cooler to hotter regions the room that is cooler will be heated. This concept was implemented by 1741, when the last model for the Franklin Stove came for sale at the marketplace, so that it could be used by homeowners in allowing more heat of their home to be able to use less amounts of burned fuel. This invention by Franklin was designed to help all people with the simplest way that was that was possible.

Chapter 11: The Death Of Benjamin Franklin

It was the well-established Pennsylvania Gazette having the black borders that became the basis of the fame and acclaim for Franklin. At the end of the day, it was that same newspaper which announced Franklin's death. Franklin. Doctor. Jones served as the medical doctor of Franklin and also provided information about the death of Franklin and all people who read The Pennsylvania Gazette. For a lengthy time, Franklin had been the victim of empyema that is a type of pus that clogged the lung. It was caused by an attack of pleurisy that afflicted Franklin several years prior to the time of his death. Franklin was experiencing a constant high temperature for a few days. It caused his breathing to be difficult, and eventually made the patient feel like he was as if he was suffocated. After enduring several days of pain and respiratory issues the patient felt more comfortable throughout the day. When he was feeling more comfortable, he wanted to leave his mattress. He pleaded with his family to tidy up his bedroom and the surrounding area as the sake of a respectable and dignified funeral. The daughter of his deceased father responded by offering a positive answer about his health and life, Franklin said he didn't think he had any optimism.

A swollen abscess was formed prior to the lung and Franklin was in the condition of incoma. It was the 17th April in 1790 that Benjamin Franklin left this world. Franklin was 84 years old at the time of his passing. old when he passed away. Two of his grandchildren Bennie as well as William Temple were at the with the founder of our nation as he lay in his final resting place.

A halt to the legacy of wisdom and invention:

On the 21st April the mourners' gatherings for the funeral procession for Franklin was conducted in the State House. A prominent clergyman associated with Philadelphia was in charge of the procession. While Franklin did not appear as a frequent churchgoer during his lifetime, he assisted many churches in arranging money to help with the maintenance and construction of the churches. The people of Philadelphia performed the funeral of the legendary Benjamin Franklin. The coffin was lit by a number of luminaries. Pall together with Philadelphia mayor at time time Samuel Powell, great Astronomer David Rittenhouse as well as a few executives from Pennsylvania. Many lawmakers and judges from Philadelphia as well as other cities as well attended the funeral. In case you were in a state of much surprise, many of printers who were from Philadelphia were also present at funerals along together with apprentices. In spite of his unique prominent position as a diplomat, he did not forget his first job in printing. He also maintained in contact with people in the business. The result was obvious at the funeral.

The other group comprised members of members of the American Philosophical Society. This was the organization to which Franklin dedicated his entire life. This was also true for fellows from the College of Physicians. Each of them was an ode to the man who established the basis for their institution, which was first medical institution in the nation. The majority of them were also from The Society of Cincinnati regardless of their disagreement with Franklin.

Here lies the father of our founding:

The cortege's wound composed of people from different walks of life were carried on until the graveyard at Christ Church. There was a report that nearly more than 20,000 mourners attended the funeral service for Benjamin Franklin. The bells of the city church were tolled. Franklin who arrived in Philadelphia as a ranaway, was being buried by the locals in a multitude. They were grieving the loss of someone who never left the world in any way from his service. The great man was interred alongside his wife, who been killed around 25 years prior. In his midst was Francis Folger, the four-year-old child of Franklin who was stricken by smallpox.

No one knew that the location that was able to provide the legendary man with a place to escape from his youth would be his final resting spot that will be a source of glory and pride.

The deceased

When we look at different areas of his career, we can see the list of achievements for Franklin is so overcrowded that there's small space left for his private life. Fans and admirers would like to know more about his private life however, very little of it is revealed. The most frequent question asked concerning Franklin concerns his kids. People want to learn what the children of his great founder father.

In the pages of the autobiography of Franklin, Franklin had three children. Three of them the one who was known as William had an unlegitimate birth because his birth was a result of an unconstitutional relationship between Franklin with a few women. The third one was Francis Folger who died at the age of just 4 years, after being a victim to the smallpox. The older daughter William Franklin lived with Franklin as did his wife.

Franklin eventually rose into the post of unwaveringly loyal governor of New Jersey. In light of the conflicting opinions of the nation's politics and its people the Governor had a complicated relation to his dad. After the Revolutionary War ended Eventually he relocated to England

The 3rd descendant to Franklin is his child, named Sarah who was also referred to as Sally by certain historians. Sarah married a man named Richard Bache and had seven children that carried the honors of their famous grandfather, however with a distinct name. Benjamin Franklin considered children to be innocent souls who require to be treated the best way possible. throughout his entire life, he held his belief in this.

The passing of Benjamin Franklin was a great loss not just for the American country, but also for all of humanity. This is due to the fact that his dynamic spirit always worked towards the well-being and prosperity of all living creatures. However, he lives within the midst of his nation, where his name is regarded as a symbol of honor and prestige.

Chapter 12: The Legacy Of Benjamin Franklin

Benjamin Franklin has survived his whole life with the influence of the renaissance spirit. Franklin was observant of the scientific and spiritual world in the world around his. Because of his curiosity, he shone shining brightly in various areas that shaped a wide range of achievements that reflected his arduous endeavors and efforts. To find a similar person in the past of America is not easy due to the sheer number of areas which he excelled in was definitely over the capabilities of a human being. Whatever he chose to do was his, he mastered it to the top quality in it.

We have a good understanding of the life of Benjamin Franklin in this publication, speaking about his achievements and presenting the details is a real matter of difficulty because describing each of his accomplishments can be a challenge. In describing the legacy left by Benjamin Franklin one word which came to mind was the certainty "diversity". Franklin was confident of his ability and expertise in so many subjects and disciplines that it amazes us in a large way. What he accomplished, the result was outstanding and power.

The most innovative scientist ever

If we think about the achievements of Benjamin Franklin as a scientist there isn't any single aspect of science as is probably the case most scientists. For him, the sky was his limit. Anything that irked him or caught his interest, he tried to study it to its highest degree. From optics to meteorology, all the way to the oceanography field, he was able to explore all of the main areas of science. As his result, he's made permanent marks of his mark in the realm of science.

His scientific pursuits were unplanned or supported by any formal training. The man was simply naturally gifted to look at the reality of existence in a manner that was different than the normal people. When something irritated him, the scientist would dig deep to discover the solution. His ability and drive to investigate, helped him be a cherished legacy an astronomer.

A diplomat quite alert

Imagine a person working with a printer and also writing his own paper before being included in the list of diplomats and ambassadors of the nascent American nation must be awe-inspiring. It was the interplay between diversity that Franklin took on. Franklin had a profound knowledge of the politics. He was a part of declaring independence, as well as several other treaties, which are regarded as a major change in the development of the American nation. So, his accomplishments as a diplomat identifies him as the founder of our nation who put in all of his effort in order to make it possible for the United Sates as one of the top nations around the planet.

The decisions he made in the time of the independence movement and revolutionary war proved to be most prudent, not just in the immediate term, but longer-term because they assisted the official government in defining an ideal international policy for the country.

An inventor much wise

Invention and science are associated heirlooms. However, the list of Franklin's inventions is lengthy. He always thought of knowledge that could benefit the people. So, he conducted experiments so to ensure that current generations along with the next generation can enjoy all the benefits of his innovations. From the bifocal glasses to the map technique, every action was designed to improve everyday life for a normal person. His expertise was not based on a return in money and he did not apply to obtain any copyrights or patents.

His ideas were incredibly beneficial because he didn't keep the secrets of science. When he was the first to learn of a scientific breakthrough, he made the information public so that all would benefit in equal measure. Perhaps it was his kindness that led him to rise to the awe-inspiring position among inventors.

Awards and honours that were earned

In addition to highlighting the many facets of the legacy of Franklin There is no reason to highlight his achievements since what Franklin earned through the way of fame and reputation through the pages of his own history surpasses any recognition or reward.

Franklin didn't win numerous awards in his career despite the nature of his things to do was long. Franklin was instrumental in establishing the first library that lends money to United States, shaped the Poor Richard's Almanac, recorded the Gulf Stream as well as published the first cartoon on politics in America and created his Declaration of Independence, became an official signer of the Constitution and also designed the glass harmonica and bifocals. The rest of his life was filled with the public image and positions that gained him in a relatively short time time. The position included the printer of official status of New Jersey and Pennsylvania and New Jersey, clerk for the Pennsylvania Assembly and The Grand Master of the Masonic Lodge in Pennsylvania Postmaster for Philadelphia and Postmaster General of North America and a representative for Congress. Continental Congress.

The year was 1753 that Benjamin Franklin was awarded the Copley Medal Form by the Royal Society of London. The award is similar in value to Nobel prize of the modern time, which was awarded to Franklin due to his remarkable contribution to electrical energy and other related sciences. It was that same year in which Franklin was awarded honorary doctorate degrees of a variety of prestigious institutions, which comprise those of Yale University of New Haven, Mass, Conn. The other was Harvard University in Cambridge. In 1759, he received an honorary doctorate in law at the University of St. Andrews in Scotland. All of these were manifestations of his abilities and unique efforts that he made in the academic circles as well as the professionals. There are still a few schools in America give their prizes the Benjamin Franklin title as an tribute to the founder father and even many years. Whatever the amount of honors may be modest but what is the acclaim that he experienced was certainly a explanation of his achievements.

Chapter 13: 48 Leadership Lessons

Establish daily routines

If you are looking to be an the most influential leader in your life You will need to develop a daily routine. Your actions should be reflective of your goals in the world. The most successful leaders have routines they adhere to regularly. If you practice a particular technique every day, it becomes an integral part of your routine. It will allow you to be more effective at job and in your daily life.

Learn useful skills

It's impossible to be successful if you don't improve your abilities over and over. Learn useful techniques and work to improve these skills. The skills you acquire are essential to making your life more productive. If you are looking to advance within your professional and personal life, it is essential to learn a new skill that can be used to your advantage within your daily life. It is an easy and successful method of ensuring an effective professional and personal life.

Create a method that allows you improve your abilities

The daily routine of your schedule must allow you to improve your capabilities. Your schedule must be designed to ensure that you are able to have plenty of time to develop your talents. It is important that you alter your schedule when you find that the program isn't efficient in your daily everyday life.

Find out how to spend your time effectively

It is a resource which must be used properly and in a correct manner. Every person has the same amount of time each day. However, what you do with your time every day is going to affect you your future. It's crucial that you recognize the significance of time. If you want to become a successful leader, you must make the most of your time with care. When you spend your time effectively on the correct things, you can make your life more enjoyable.

Make sure you allocate your time to the correct actions

You'll have a lot of tasks which you will need to accomplish in one day. There are many tasks that require your attention. If you want to be effective in your job through being a successful manager, you'll need be able to dedicate the appropriate amount of time for the appropriate tasks. It is something you must work on. It is important to ensure that your most important and urgent assignments are given the appropriate quantity of time.

Work on developing purposeful routine

An entrepreneur will always be able to find a reason within his head when he plans the day. Your schedule for the day must be in a position to give justice to the goals you've set out to accomplish. Someone who is focused on having a well-planned routine is constantly productive at his job. Make sure every day time isn't wasted and used correctly and the correct direction.

Prioritizing your tasks

There are times when you have to fulfill several commitments during the course of a day. A great way to meet every obligation and accomplish all the tasks you have to do is to prioritize your work carefully. It is important to determine which projects must be finished first, and which tasks are best left to be put off.

Try your art in silence

If you are looking to live an enjoyable and prosperous life it is important to understand what needs your focus. It's important to be able choose when and where you should speak, and when it's time to remain silent. The most successful leader can use silence for the benefit of others. Avoid the temptation to talk about all the happenings within you. It is best to only speak when absolutely necessary.

You must follow an instruction for items you perform

If you are looking to become successful in whatever you are doing in your life, you have to be aware of what to accomplish and the time you must complete it. It is essential to establish a plan to accomplish the work that needs to be accomplished. A simple practice can be a big help in helping you be productive in your tasks. It is possible to establish up a schedule for your routine tasks that you must to finish.

Leaders will complete the work in the proper time.

There is a need to find a solution

An effective and successful leader recognizes that they will have to make resolutions time and time to time. Make these resolutions element of your routine. They can help you improve your performance. If you are looking to become an effective leader, take a look at your day-to-day activities and create daily resolutions. It is an easy method to become effective every aspect of your daily life.

There is a need to create your agenda for the day.

It is essential to create an agenda for your day. Your agenda must reflect the purpose of your day. This is a task that must do first thing to do when you wake up. There are numerous tasks can be completed throughout the day, but there must be a goal for why you're working. The key is to define the goal and plan to ensure that you are able to have a purpose that drives your daily activities and throughout your day.

Review your work

It is a good idea to make it a routine to take a moment and reflect on the work you've done. It's important to do this to be sure that things have been progressing in the manner you intended they would. Do not be a slave to your work without reflecting and review of your work. Everyone who is successful makes their routine a habit of daily review everything they accomplish during their working day.

Chapter 14: You Must Be Critical About Your Work

If you want to be successful in your career as well as in everyday life, then you must examine your work with a critical eye. It is crucial to look at your work in a way that is objective. This is an attribute that all great leaders possess. The leader is always conscious of his actions. Your approach to work helps you improve both as a leader and also as an individual leader.

be humble

It is important to practice humility as an integral part of your everyday routine. Simple acts of humility can change your life to an incredible amount. A lot of successful individuals consider humbleness as a key factor in their success. Your life can be transformed to the best possible ways by practicing humility, not only in your job, but in your daily life. An effective leader or businessperson can demonstrate humility, along with numerous other virtues essential to success.

Cleanliness is a must.

If you are looking to increase the standard of your life it is essential to take up the practice of keeping your surroundings clean. Cleanliness is something you must practice both in your thoughts and actions. Also, you must ensure that the environment is clean. One who has pure thoughts, behaviors and environment is one who is averse to any thing that may cause harm throughout his life. It's a straightforward method of improving the quality of your life in general.

Avoid extremes.

If you are looking to make improvements in your professional life as well as your overall life You must know the proper quantity of work you must put into any endeavor. Successful people will be cautious about anything that's extremely difficult. If you're incredibly inactive, it isn't suitable for your health. If you are also over your work hours, it might be detrimental to your health. It is essential to learn the importance of moderation and stay clear of the extremes of all sorts.

Make sure you are thinking rationally and efficiently.

The most successful leaders will consider the facts. He'll know what's essential and not. It is not possible to let your thoughts or your heart rule your life. You must be consistent when you make decisions. It is crucial if you look at easy but successful ways to improve the quality of your life. It is important to always seek out ways to make your life better as well as your job.

Be honest and work hard

If you are looking to become prosperous in your career it is essential to be an honest person. It is essential show sincerity in all you undertake. Make sure you're sincere about the work you do and in your daily everyday life. Someone who pursues his goals with integrity will always be ahead of the rest. Make sure you are honest every day.

Recognize the importance of moderate consumption

If you want to emulate the traits of a successful person in your own life, then you must understand the significance of moderation throughout your day. The importance of the importance of moderation. If you don't take things that are too extreme You create a balanced framework for your life. It will allow you to be more successful in your everyday routine. You'll be able to take note of everything every day brings.

It is important to always possess an awareness of justice

If you're looking to making improvements to the level of your living It is essential to maintain an attitude of fairness in your work and life. You must be able discern the good from the negative and distinguish the significant from the nebulous. Someone with an innate sense of justice always successful in what he decides to pursue in his life. The simple act of justice can make a difference throughout your lifetime.

Take away the things that aren't important.

If you are looking to succeed at your field of study as well as in the rest of your life, you must decide the most important things for your job and your development. It is essential to eliminate out the items that aren't crucial and eliminate the items do not have any significance for your progress. It is a process that continues which means you'll be required to repeat it frequently. This can help you increase your quality of life as well as your workplace culture. It will be evident that the majority of succeeding people use this method.

Stay focused

Everyone who has succeeded in their careers and lives have learned to remain focused on the work that they have to complete. Yet, even accomplished people have to deal with a variety of problems, like staying focused. There is how to stay focused in your job. This can help you enhance your lifestyle in general. Always look for ways to be focused at work.

Accept the challenge

If you are looking to achieve excellence at what you do then you must acknowledge that it will not be easily. It is not possible to get something on a plate. If you can accept your struggle and accept the challenge, it will be more manageable to you to accept the challenge. This helps you stay focused and inspire you to strive harder for the end goal. The most effective way to be successful is to embrace the obstacles which are aspect of the process.

Create a daily score card

Be critical of the progress you make and how well you are doing. This is an easy method to improve your personal life. It is recommended to keep your daily scorecard in order to keep track of the progress you make. Your daily score sheet will help you analyze your efforts more effectively. Look for areas that need improvement or improvements. You should create a few columns on your scorecard. They are likely to allow you keep track of the progress you have made.

Keep track of your daily progress

Everyone who is successful evaluates their performance regularly. time and time. It is possible to create the road map you want to follow and then proceed in accordance with to it. If you fail to keep track of your progress, there is no point in all that hard work. It is essential to establish goals to yourself on a regular basis. Also, you should be able monitor your performance on a regular basis. This can help you increase the areas where you're lacking.

Chapter 15: Change Your Plan

It is essential to keep track of your progress on a regular basis. If you keep track of your progress, you will be able to recognize those aspects that aren't effective for you. Make the necessary changes to your routine as well as your time table. This will allow you to enhance the quality of your job and daily life.

Be aware of the importance of organisation

The importance organization in your daily day-to-day life. Organised people can achieve far more than someone who's not organised. It is important to make organization an integral part of your routine. If you do it consistently and regularly, it becomes an integral part of your routine.

Don't forget the most important things

It is essential to be a smart and savvy enough to understand the things that matter to you. It is likely that you will encounter many difficulties and interruptions throughout your day everyday life. However, it's your obligation to concentrate only on things that are essential to you. Whatever you're doing, it is important to be clear on the purpose on your mind.

You must adhere to the strict routine

If you want to enhance the way that you live your life, then you have to adhere to a certain routine. This routine can help your brain become more better trained, and also assist you to concentrate more efficiently.

Do not wait around until inspiration strikes

It is not a good idea to sit and expect idea to come into your mind. The way it works is not that simple. Actually, you must remain focused on your timetable. It is likely that you will get inspiration when you are working. However, you shouldn't lose time sitting around waiting for a magical idea to appear by itself.

Enjoy the little things in your daily

Someone who is determined to succeed in all areas in his life must look for joy in the small aspects of everyday life. Everyday items with gratitude and love. When you view the little things in your life with love it will bring you immense happiness and joy within your own life. This can help you have a peaceful and happy life. tranquil.

Enjoy spending time with your family

Your family is the most precious thing you have. Don't become so occupied that you can't make time to spend time with your loved ones. Whatever your schedule are, make sure you take time out of your schedule to take time to be with your loved ones. The time that you spend with your loved ones can inspire you to become more successful in all aspect.

You should take time to do things you enjoy

Always make time to indulge in your passions throughout the course of your daily life. Spend time doing things that you love doing. This is a great approach to refresh your spirit. If you're calm and calm, you do more effectively in every area that you are in. It will be evident improvements at work.

Thank God for all that you've got.

It's vital to be content with what you have within your daily the world. As you strive to be more successful things in your life, it is important to make the time to reflect on your present circumstances. One should not forget the joys he has within his own life.

Put money into good relationships

The quality of your life is enhanced if you have relationships that are healthy. It is important to spend time with your family and friends who are understanding of you. It is important to work towards building your bonds to them greater. Always try to spend your valuable time with people who are meaningful to you only.

Do not let the pressure pull your down.

It is likely that you will encounter many difficulties every day. They will cause anxiety as well. Learn how to manage anxiety effectively. Always keep in mind that you must not allow stress to take control of your life. It is important to remain confident.

Take time to unwind

It is essential to relax to a person. While it's essential to be a hard worker however, it's also essential to make time to unwind. There are many ways to relax. an active part in things that allow your body unwind. The concept behind relaxing time is to completely get rid of all everyday activities and be at peace.

Be patient and don't eat your food too fast.

There are many who consider eating food as a duty to fulfill. Take the time to relax and enjoy food. It is essential and can aid in reviving your body. Be sure to give adequate time for meals into your day-to-day schedule.

Pause for a break

Also, you should include plenty of breaks throughout your day. It's essential to get your work done at a steady pace, however it could reduce productivity. Make breaks and break times in your routine and take advantage of it to the max.

Be aware of when you should walk away

Follow a strict timetable. However, you need to recognize when to take a break and walk off. time off. Sometimes, you get stuck at job. You should take a break and unwind of your job. This can help you feel more energized and enable you improve your performance.

Spend time to learn

It is essential to learn at all times in every day life. Whatever you are aware of and how much you aren't aware of it is important to invest your time to learn. This can help you develop in your career. When you spend your time studying it is investing your time and energy into the future of your life. Always learn something new in your lifetime.

Never give up

Never quit, no matter if you're failing, or in the event that you do not know what. If you're not confident in your understanding, then it is best to learn more however, you should never quit. Always keep your spirit high and do your best in achieving what you would like to accomplish.

The focus should be on leadership within your own personal

If you are looking to make positive improvements in your lifestyle it is essential to concentrate on your personal leadership. When you're able to exhibit the traits of personal management in your daily life and you'll see positive change in your lifestyle. Make sure that you are paying close attention to the daily routines you follow. Being more aware will allow you become better at everything you do.

Chapter 16: Finish Your Day In A Positive Way

It's important to enjoy your day to the maximum and finish your day on an optimistic way. You, as a person, must strive to be the best you can to every day that is ahead. Once you're done with the day then you ought to be able to close the day with a smile and thankful note. The simple act of being thankful will allow you succeed throughout the day.

Examine the tasks that comprise your day

It is a good idea be able to go over your goals for your day prior to going to sleep. It is crucial since you'll be able track the progress you have made. Be committed to your goals. This can help you recognize your strengths as well as weak areas. It is recommended that you review this every evening before going to sleep.

Review your work

It is essential to take a moment to reflect on the progress you have made towards the end of every day. You must be able to identify the things that would have been more effective. It is also important to thank your self for the hard work you have done. It is important to perform this easy routine at the conclusion of every day. This can help you keep track of your progress.

Think about what you did

If you want to live an enjoyable life and have a fulfilling career, you must contribute every day towards your larger goals. Prior to going to sleep each night, consider the good you did this day. It is crucial and can help you help you achieve your goals to achieve in your your life.

Rest soundly

It is important to never undervalue the significance of sleep to your daily life. If you're looking to making improvements to your quality of life it is essential to pay attention to your hours of sleep. When you're done with the daily routine and review your entire day, it is important to be looking forward to sleeping soundly. It is essential to incorporate at minimum six to eight hours sleep throughout your day.